ISOTOPES IN ORGANIC CHEMISTRY

VOLUME 6
ISOTOPIC EFFECTS:
RECENT DEVELOPMENTS
IN THEORY AND EXPERIMENT

ISOTOPES IN ORGANIC CHEMISTRY

Edited by

E. BUNCEL

Queen's University, Kingston, Ontario, Canada

and

C.C. LEE

University of Saskatchewan, Saskatoon, Saskatchewan, Canada

VOLUME 6
ISOTOPIC EFFECTS:
RECENT DEVELOPMENTS
IN THEORY AND EXPERIMENT

ELSEVIER

Amsterdam — Oxford — New York — Tokyo   1984

ELSEVIER SCIENCE PUBLISHERS B.V.
Molenwerf 1
P.O. Box 211, 1000 AE Amsterdam, The Netherlands

Distributors for the United States and Canada:

ELSEVIER SCIENCE PUBLISHING COMPANY INC.
52, Vanderbilt Avenue
New York, N.Y. 10017

Library of Congress Cataloging in Publication Data
Main entry under title:

Isotopic effects.

   (Isotopes in organic chemistry ; v. 6)
   Includes bibliographies and index.
   Contents: Isotope effects on ^{13}C NMR shifts and coup-
ling constants / by David A. Forsyth -- The effect of
pressure on kinetic isotope effects / by Neil S. Isaacs
-- Magnetic isotope effects / by Nichols J. Turro and
Bernhard Kraeutler -- [etc.]
   1. Chemical reaction, Rate of--Addresses, essays,
lectures. 2. Isotopes--Addresses, essays, lectures.
I. Buncel, E., 1931-   . Lee, C.C. (Choi Chuck),
1924- . III. Series.
QD502.I76 1984   547.1'39   84-8132
ISBN 0-444-42368-0 (U.S.)

ISBN 0-444-42368-0 (Vol. 6)
ISBN 0-444-41742-7 (Series)

Printed in The Netherlands

ISOTOPES IN ORGANIC CHEMISTRY

Contributors to Volume 6

David A. Forsyth

Chemistry Department
Northeastern University
Boston, Massachusetts, U.S.A.

Neil S. Isaacs

Chemistry Department
University of Reading
Reading, England

Bernhard Kraeutler

Laboratorium fur Organische Chemie
Eidgenossische Technische Hochschule
Zurich, Switzerland

David E. Lewis

Department of Chemistry
Baylor University
Waco, Texas 76798, U.S.A.

Leslie B. Sims

Department of Chemistry
North Carolina State University
Raleigh, North Carolina, U.S.A.

Nicholas J. Turro

Chemistry Department
Columbia University
New York, New York 10027, U.S.A.

ISOTOPES IN ORGANIC CHEMISTRY

Volume 1. Isotopes in molecular rearrangements

N.C. Deno The Pennsylvania State University	Deuterium labeling in carbonium ion rearrangements
W.R. Dolbier, Jr. University of Florida	Isotope effects in pericyclic reactions
J.L. Holmes University of Ottawa	The elucidation of mass spectral fragmentation mechanisms by isotopic labeling
D.H. Hunter University of Western Ontario	Isotopes in carbanion rearrangements
J.S. Swenton The Ohio State University	Utilization of deuterium labeling in organic photochemical rearrangements

Volume 2. Isotopes in hydrogen transfer processes

M.M. Kreevoy University of Minnesota	The effect of structure on isotope effects in proton transfer reactions
G. Lamaty Université de Montpellier	Deuterium exchange in carbonyl compounds
K.T. Leffek Dalhousie University	Proton transfers in nitro compounds
E.S. Lewis Rice University	Isotope effects in hydrogen atom transfer reactions
H. Simon and A. Kraus Technische Universität Munich	Hydrogen isotope transfer in biological processes
P.J. Smith University of Saskatchewan	Isotope effects in elimination reactions
R. Stewart University of British Columbia	Isotopes in oxidation reactions

Volume 3.   Carbon-13 in Organic Chemistry

G.E. Dunn University of Manitoba	Carbon-13 kinetic isotope effects in decarboxylation
J. Hinton, M. Oka and A. Fry University of Arkansas	Carbon-13 n.m.r. methodology and mechanistic applications
G. Kunesch and C. Poupat Institute de Chimie des Substances Naturelles du C.N.R.S.	Biosynthetic studies using carbon-13 enriched precursors
A.S. Perlin McGill University	Application of carbon-13 n.m.r. to problems of stereochemistry
A.V. Willi Universität Hamburg	Kinetic carbon and other isotope effects in cleavage and formation of bonds to carbon

Volume 4.   Tritium in Organic Chemistry

J.A. Elvidge and J.R. Jones University of Surrey and V.M.A. Chambers and E.A. Evans The Radiochemical Centre Amersham	Tritium nuclear magnetic resonance spectroscopy
W.J. Spillane University College, Galway	The use of tritium and deuterium in photochemical electrophilic aromatic substitution
Y.-N. Tang Texas A&M University	Reactions of energetic tritium atoms with organic compounds
D.W. Young University of Sussex	Stereospecific synthesis of tritium labelled organic compounds using chemical and biological methods

Volume 5.   Isotopes in Cationic Reactions

C.C. Lee University of Saskatchewan	Degenerate rearrangements in triarylvinyl cations
S. Oae and T. Numata University of Tsukuba	Pummerer rearrangements
R.M. Roberts and T.L. Gibson University of Texas at Austin	Friedel-Crafts reactions: some isotopic studies
D.L.H. Williams Durham University and E. Buncel Queen's University	Aromatic cationic reactions

FOREWORD TO VOLUME 1

Organic chemistry is characterized by a vast variety of compounds, structures and reactions realized by a rather limited number of chemical elements. One and the same element is generally represented by a considerable number of atoms, playing several different roles. It is evident that a method enabling us to give the otherwise anonymous atom a kind of identity should be of particular value in this branch of chemistry.

Tracing by means of similar but still chemically discernible groups has been practised in organic chemistry for a long time, and has revealed that organic reactions are far more varied than expected. Isotopes, being chemically identical in a qualitative and usually also in an almost quantitative sense, are far more powerful as tracers, due to this similarity and to the fact that atoms rather than groups are labelled and can be traced as such.

A simple account of the molecular species involved, their structures and configurations, cannot be considered a complete description of a chemical system in equilibrium. We know from studies of non-equilibrium systems that opposite reactions balancing one another are generally taking place. As far as species of different molecular compositions are concerned, this has been realized for more than a century. It is only in the last few decades, however, that we have had at hand the means to measure the amounts of different isotopes and follow the behaviour of systems which are non-equilibrium ones with respect to isotopic composition. This has led to a still more vivid picture of most systems in equilibrium, with several exchange reactions taking place, sometimes at rates too large to be measured on the classical time scale of chemical reactions.

Even this is not enough for the true scientist who wants to go beyond the knowledge of which reactions actually take place and how fast they occur. He feels a desire to know also how the atomic nuclei and electrons behave in the transition called a chemical reaction. His questions come close to the fundamental limit set by the principle of uncertainty. At present the transition state of the rate-determining reaction step seems to be the most complete description attainable. In such studies it is not only the qualitative chemical similarity of isotopes, allowing the identity of atoms in the transition state to be revealed, but also their quantitative chemical dissimilarity which is of importance and allows a study of the force field and hence binding conditions in the transition state itself. Thanks to the fairly low atomic number of most atomic species of importance in organic chemistry, the relative mass differences between isotopes are sufficient to cause differences in quantitative behaviour, rather easily measurable with modern instruments.

Many scientists feel the flood of scientific publications as an encumbrance. The justification for the existence of a series like the one started by the present volume lies in the aid that the surveys it contains may offer the research worker and, perhaps more important, the stimulus for further research that may be provided. The application of isotope methods has undoubtedly a very important role in future research in organic chemistry. No attempt at a detailed prediction will be ventured here, however. It may suffice to refer to the development of the nuclear magnetic resonance technique. The studies of ordinary hydrogen nuclei, which have been of outstanding importance for the development of organic chemistry, can be considered as an application of isotope methods according to the ordinary usage of the concept only to the extent that deuterium has been used as a stand-in for protium. In the not-too-distant future, however, most laboratories will have equipment allowing routine studies of the less abundant carbon isotope ^{13}C, and then many chemists will be in possession of a sensitive probe in the centre of atoms of the most important element in organic chemistry. It will reveal not only details of molecular structure in the usual sense but also more subtle details about the electron distribution in the backbone of organic molecules. It is open to discussion, of course, whether this kind of work, which frequently makes use of the natural occurrence of heavy carbon, should be considered as an application of isotope methods. In any case, it utilizes a particular property of an isotope different from the most abundant one.

It is evident from the thoughts expressed in the last paragraph that the borders of the field "Isotopes in Organic Chemistry" are rather indeterminate. The editors' intention to apply as few restrictions as possible on subject matter seems wise, because then the interest taken in the present series by its future readers can be allowed to indicate the position of these borders in practice.

Göteborg   Lars Melander

CONTENTS

Chapter 1

ISOTOPE EFFECTS ON ^{13}C NMR SHIFTS AND COUPLING CONSTANTS

DAVID A. FORSYTH

Chemistry Department, Northeastern University,

Boston, Massachusetts

CONTENTS

I.   INTRODUCTION

(A)   Isotope Effects on NMR Spectra

    Isotopic substitution can affect the nuclear magnetic resonance spectra of organic molecules in several ways and has become a popular means of extending the usefulness of the nmr method.   One source of the effects is the difference in the magnetic properties between isotopes.   A source of more subtle effects is the difference in mass between isotopes.   The latter type is analogous to isotope effects on rates and equilibria in having fundamentally a vibrational origin, while the former is perhaps more akin to the occurrence of radioactive decay for some isotopes.   This review focuses on isotope effects on nmr chemical shifts (the isotope shift) and reduced coupling constants which are associated with isotopic mass differences.   The emphasis in this review is the observation of isotope shifts of ^{13}C resonances and isotope effects on spin-spin coupling to carbon, though isotope effects on other nuclei and other couplings are discussed to provide perspective.

    Although not discussed in detail here, changes in nmr spectra due to changes in magnetic properties upon isotopic substitution are extremely useful in nmr studies.   The change in magnetogyric moment has long been used in proton nmr, where substitution by deuterium removes the signals of the replaced protons from the spectrum due to the change in Larmor frequency and also effectively collapses the multiplet patterns due to much smaller coupling to the deuterium nuclei.   In a similar vein, ^{13}C enrichment increases the signal intensity for the labelled carbons in a ^{13}C spectrum and splits resonances of coupled protons in a ^{1}H spectrum.

The magnetic effects of deuterium substitution also produce valuable effects in ^{13}C spectra, most notably a reduction in signal-to-noise ratio for the deuterated carbon to the extent that the signal may be effectively removed from the spectrum.[1] The effective signal-to-noise ratio is reduced for several reasons: (i) Transverse relaxation (T_2) due to the deuterium quadrupole moment may broaden the ^{13}C signal. (ii) In a ^{1}H-decoupled spectrum the deuterated carbon still appears as a multiplet due to one-bond ^{13}C-^{2}H coupling.[2] (iii) If a carbon is completely deuterated or if it has no attached hydrogen atoms but is close to the site of isotopic substitution, the nuclear Overhauser enhancement (NOE) will be reduced in a ^{1}H-decoupled spectrum.[3] (iv) Deuteration may lengthen spin-lattice relaxation times (T_1) due to less effective relaxation associated with the smaller magnetic moment.[3-5] Another useful effect of deuteration in some cases is two- or three-bond coupling to deuterium, which may give broadened ^{13}C resonances or resolved splitting occasionally.[3] In a proton-coupled spectrum, deuteration will alter the multiplet patterns.

The mass-related effects of isotopic substitution on nmr spectra are also receiving attention as an aid to signal assignment and as a potential source of detailed structural information. The isotope effects on nmr parameters may be classified as primary or secondary in nature. Primary effects occur when the isotopic substitution is at the nucleus whose shielding is of interest or at a nucleus directly involved in the coupling of interest.[6] For instance, primary isotope shifts would be found by comparing ^{1}H, ^{2}H, or ^{3}H chemical shifts in ppm for a particular substance, referenced to the appropriate protonated, deuterated or tritiated standard, or by comparing peak separations. These isotope effects are ordinarily very small,[7] except in the case of systems with relatively strong hydrogen bonds.[8-12] Obviously the primary isotope effect can not be detected for the carbon series ^{12}C/^{13}C/^{14}C, because only ^{13}C has a magnetic moment. Secondary isotope effects are those induced in a chemical shift by substitution at another nucleus or in a coupling between two nuclei by substitution at a third nucleus. Secondary isotope effects on ^{13}C chemical shifts have been of most interest in terms of practical applications because isotope shifts can help identify not only the immediate site of isotopic substitution but also nearby positions.

4

An example of the effects of deuterium substitution on ^{13}C spectra is shown in FIGURE 1, which displays ^{13}C spectra for benzo[c]phenanthrene and benzo[c]phenanthrene-2-$\underline{d}$.[13]   The deuterated material gives separate signals for C_1 and C_{12}, and for C_3 and C_{10}, whereas chemical shifts are equivalent within these pairs in the unlabelled compound. C_1 and C_3 are shifted upfield by the isotope shift. C_2 appears as a weak triplet, shifted upfield of its former position coincident with the C_{11} signal. The C_4 and C_{12a} signals are broadened due to three-bond coupling, J_{CCCD}.

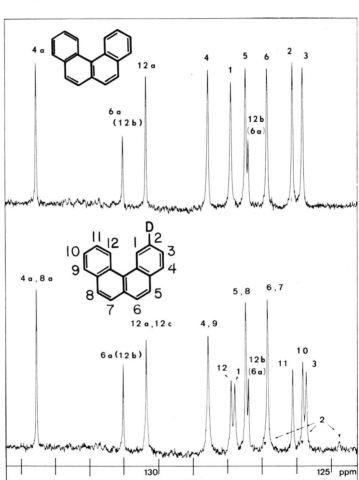

FIGURE 1. ^{1}H-Decoupled ^{13}C NMR spectra of benzo[c]phenanthrene and benzo[c]phenanthrene-2-$\underline{d}$.   Reproduced by permission of Pergamon Press.[13]

A distinction must be made between isotope effects which are intrinsic, i.e., characteristic of an isolated, "static" molecule, and equilibrium isotope effects.[14] Equilibrium isotope effects on nmr parameters result when the position of an equilibrium in a process which is fast on the nmr time scale is perturbed by isotopic substitution. Measurement of equilibrium isotope effects is a powerful new method to study fast rearrangements or conformational changes.[15] The distinction between intrinsic and nonintrinsic effects may be quite difficult when it is considered that isotopic substitution may also affect hydrogen bonding and other forces of solvation and intermolecular association. It is reassuring to this author that at least methane displays isotope effects on the chemical shifts and coupling constants in the gas phase.[16,17]

Many observations of isotope effects on ^{1}H and ^{19}F spectra as well as theories of the isotope shift were discussed in a review by Batiz-Hernandez and Bernheim that appeared in 1967.[18] Stothers commented on isotope effects and their applications in ^{13}C nmr spectroscopy in a 1974 review.[3] Wehrli and Wirthlin described the use of the isotope shift for ^{13}C signal assignment in their 1976 text.[19] Of course, most research reports refer to earlier work, but a briefly annotated and fairly extensive list of references was provided by Risley and Van Etten in a 1980 paper on ^{18}O isotope effects in ^{13}C nmr.[20] The references in this latter paper include studies of isotope shifts in a variety of nuclei which are not reviewed here.

Judging from the isotope effects on ^{13}C nmr spectra described in the sections that follow, the conclusion of Batiz-Hernandez and Bernheim is still valid that isotope shifts are unlikely to provide detailed information about a particular structural parameter because of the variety of influences on the isotope shift.[18] However, the isotope effect on nmr properties is an interesting phenomenon which now has associated with it many actual and potential applications. The occurrence and applications of intrinsic isotope effects in ^{13}C nmr spectroscopy are reviewed here.

In referring to isotope shifts for ^{13}C nuclei, the effect of substituting a heavier isotope will be defined by eq. 1.

$$\Delta \delta_{c} = \delta_{c}(\text{undeuterated}) - \delta_{c}(\text{deuterated}) \qquad (1)$$

The isotope shift, $\Delta\delta_C$ in ppm, is defined in this way so that upfield shifts, which are typical, are positive values. Isotopic substitution involving a nucleus other than deuterium is indicated by writing the symbol of the heavier nucleus in parentheses, for instance, $\Delta\delta_C(^{18}O)$ for $^{16}O/^{18}O$ isotopic substitution.

B.   Theoretical Concepts

The isotope shift is fundamentally vibrational in origin, since isotopic substitution does not affect the electronic potential energy surface in a molecule. For a fixed nuclear configuration, the distribution of electrons depends on nuclear charges, not nuclear masses (Born-Oppenheimer approximation). However, the effective geometry and average electronic structure for isotopically substituted molecules will differ, and hence nmr shielding will differ, because the dynamic nature of the molecule must be considered. The observed nmr shielding is not for a fixed nuclear configuration, rather it is the average shielding over the contributing vibrational and rotational states and such states are dependent on nuclear masses. Most theoretical treatments of the isotope shift, whether formal or qualitative, deal with the phenomenon in terms of the effective change in molecular geometry and molecular wavefunctions due to vibrational averaging.

The formalism of the theory of the nmr isotope shift is closely related to the theory of the temperature dependence of chemical shielding.[21-23] The contributions to the temperature dependence or isotope effects come from anharmonic vibration, centrifugal distortion, and a small contribution from harmonic vibration.[23,24] A complete theory of the isotope shift requires knowledge of all the force constants in contributing vibrational and rotational states as well as accurate electronic wavefunctions for calculation of shielding. It has been suggested that since the isotope shift is vibrational in origin, it should be temperature dependent and approach zero at high temperatures as higher vibrational levels become populated.[25,26] An observation that the temperature dependence of the two-bond isotope effect of $^{35}Cl/^{37}Cl$ substitution on the ^{19}F shift in $CFCl_3$ was more temperature dependent than the one-bond effect of $^{12}C/^{13}C$ substitution was explained by the suggestion that two-bond effects were expected to be associated with bending vibrations which occur at more easily excited, lower frequencies than the stretching vibrations that are expected to be more important for one-bond effects.[25]

Ramsey first predicted that nmr isotope shifts would be observed due to differences in zero-point vibrational functions which result from different isotopic masses.[27] Marshall calculated functions for the dependence of the screening constant on interatomic separation in the hydrogen molecule and used these functions to calculate average screening constants for H_2, HD, and D_2 assuming the ground vibrational state.[28] Saika and Narumi included both vibrational and rotational effects but used a function for the bond length dependence of shielding in the hydrogen molecule that was obtained empirically.[29] Raynes also included the effect of centrifugal distortion when dealing with the temperature variation of the chemical shift, isotope shifts, and spin-spin coupling in H_2 and its isotopomers (molecules differing only in their isotopic composition).[6,22] Raynes used ab initio calculations to determine the dependence of shielding on bond lengths. Raynes also calculated isotope shifts for isotopomers of carbon monoxide.[26] In discussing the theory of isotope effects, Buckingham and Urland noted that "the harmonic and cubic anharmonic force constants and the equilibrium value and first and second derivatives of the property with respect to the internal coordinates must be known" in order to predict values of a physical property for isotopomers.[24] Thus, accurate treatments of the isotope shift are likely to be restricted to diatomics and small polyatomics although it was suggested that the effect of deuterium could be approximately described by treating vibrations localized in the bond containing it.[24]

For substitution by deuterium, X-D bonds have lower zero point vibrational energies and smaller vibrational amplitudes than X-H bonds. Theoretically, isotope effects on nmr shifts may occur even if vibrations are harmonic,[28,30] but a recent theoretical treatment by Jameson suggests that the chief contribution to isotope shifts comes from the anharmonic part of the expression for nmr chemical shielding.[23] If vibrations are anharmonic, the average X-D bond will be shorter than X-H. Experimental studies indicate that bonds to deuterium are slightly shorter (for example, C-D bonds[31,32]). The effect of bond shortening is increased shielding of the two nuclei involved.[23] In an important early study, the effect of ^{13}C substitution for ^{12}C on the [54]Co chemical shift of $K_3Co(CN)_6$ was found to be in the same direction as the effect produced by an increase in pressure, which would be expected to give a shortened Co-C bond.[21,33] Sergeyev calculated isotope shifts in some small molecules such as methane by simply assuming a C-D bond is 0.01Å shorter than a C-H bond.[34,35] Shielding differences were then estimated by calculating para- magnetic shielding for the two geometries by using double pertur- bation theory. Sergeyev predicted a 0.17 to 0.60 ppm shielding for CH_3D, depending on his method of approximation;[35] the observed shielding is 0.19 ppm.[17]

Bernheim and Batiz-Hernandez introduced the concept of a change in hybridization of the central atom upon substitution by deuterium to explain geminal (H-X-H) isotope shifts in ^{1}H nmr.[16,18,36] The rehybridization concept has also been applied to isotope effects on coupling constants.[37,38] It is reasoned that there will be more s character in the hybrid orbital of X used for bonding to deuterium because of the effectively shorter bond to deuterium. The rehybridization will result in different bond angles and less s character in the remaining bonds. Together, the changes in bond lengths and bond angles produce the shielding and coupling changes.

However, rehybridization cannot play a role in certain molecules of high symmetry if isotopic substitution is made symmetrically. Jameson pointed out that in molecules of the CO_2, CH_4, BF_3, or SF_6 types of geometry, any isotope shift must come from anharmonic vibration or second order harmonic terms; the rotational (or centrifugal distortion) contribution to the isotope shift is zero.[23] In regard to hybridization, exact sp^3 hybridization must apply to CH_4 and CD_4, although not to partially deuterated species. It is interesting that the isotope shift for ^{13}C in the series $CH_4/CH_3D/CH_2D_2/CHD_3/CD_4$ is almost precisely additive and linear,[17] while the isotope shift for ^{1}H in the series is not additive.[16] Hybridization changes obviously do not influence the ^{13}C shift, but could be a factor in the ^{1}H shifts. In a similar series, a nonlinear dependence of the [119]Sn chemical shift on the number of CD_3 groups in tetramethyltin was found.[39] The deviation from additivity was interpreted as indicating a contribution from centrifugal distortion, although the effect of anharmonic vibration was considered to be dominant.[39]

The generalization that different average electronic wavefunctions apply to molecules differing in their isotopic composition is broad enough to cover any isotope effects on nmr parameters. Ordinarily the isotope effect is quite localized, affecting most strongly the directly attached atoms. This suggests that electron redistributions are usually quite localized. Servis has pointed out the analogy to inductive effects,[40] which also rapidly attenuate with the number of intervening bonds, although inductive electronic effects arise from electronegativity differences that are not relevant to isotope effects. However, if there are bonding interactions with the C-H(D) bond, these interactions may lead to longer range or stereochemically dependent isotope shifts which are analogous to resonance effects in their mode of transmission.[40] Such interactions were first invoked in the form of vicinal delocalization of σ electrons to explain the stereochemical dependence of vicinal H(D)-C-C-F isotope effects.[41,42] A similar explanation, in terms of hyperconjugative interactions with C-H(D) bonds, must be invoked to explain long-range and downfield shifts in carbocations[40,43,44] (see SECTION IV).

The isotope shifts produced by deuteration at groups attached to ionic π-systems[40,43,44] are analogous to secondary alkyl β-deuterium isotope effects on rates and equilibria. Hyperconjugation is generally accepted as the major source of secondary β-deuterium isotope effects on rates of reaction involving carbocation-like transition states and on equilibria involving carbocations.[45-47] Similarly, secondary β-deuterium isotope effects have been observed for reactions involving negative ions and lone pairs, and have been cited as evidence for negative hyperconjugation.[48,49] In nmr studies of stable cations, the CH_3 group appears to be a better electron donor than the CD_3 group since the deuterated ions exhibit deshielding.[40,43,44] In anions, or when attached to atoms with lone-pairs, the CH_3 group appears to be a better electron acceptor than the CD_3 group. These nmr isotope effects mimic the direction of the known kinetic isotope effects as though electronic substituent effects were responsible for the reactivity differences. However, it is the zero-point vibrational energy differences rather than the energy differences associated with charge redistribution due to anharmonicity that account for the kinetic isotope effects.

Both the kinetic isotope effects and the charge distribution differences inferred from the nmr isotope shifts are reflections of the hyperconjugative weakening of the β C-H bond. In contrast to kinetic isotope effects, isotope effects on nmr shifts involve the properties of only a single species, not a change between ground state and transition state. The observations of isotope shifts at positions remote from the site of isotopic substitution in conjugated ions (SECTION IV) are indicative of differences in electron distribution produced by the isotopic substitution. These apparent "electronic" effects arise because the effectively shorter, stronger C-D bond (effectively lower lying σ bonding orbital) will be less able to interact with an adjacent empty p orbital, which will result in reduced electron delocalization to the cationic π system.[40,44] In the same fashion, a shorter C-D bond will have an effectively higher lying σ* antibonding orbital which will be less capable of accepting electron density from an adjacent filled orbital in an anion or lone-pair system. Thus, even though the same potential energy surface applies to each ion in an isotopically related series, the ions may all have different average π electron distributions that fundamentally arise from vibronic differences rather than ordinary substituent electronic effects.

The 1967 review of the isotope shifts by Batiz-Hernandez and Bernheim examined the development of theories to account for the isotope shift and concluded that the effect was fairly well understood for the hydrogen molecule but not for polyatomic molecules.[18] While the general outline of the theory needed to account for isotope effects on nmr properties has been formulated, the capability of making accurate predictions for complex molecules is not available.[23,24] The sensible approach would appear to be to find empirical generalizations and correlations which could stimulate the development of approximate theoretical methods, much as generalizations made by Batiz-Hernandez and Bernheim[18] were the focus of a theoretical discussion by Jameson.[23] The state of empirical knowledge, through 1980, about intrinsic isotope effects on ^{13}C nmr parameters is reviewed in the following SECTIONS II-V, and some of the possible generalizations are briefly summarized in the concluding SECTION VI.

II. ONE BOND ISOTOPE EFFECTS ON ^{13}C SHIFTS

(A) Deuterium-induced Isotope Shift

The one-bond or α-effect of deuterium substitution on ^{13}C chemical shifts is expected to be the largest of possible isotope effects on ^{13}C shifts, with the exception of α-tritium substitution.[23] The relative change in mass from ^{1}H to ^{2}H is greater than is encountered among isotopes of other elements. Also, isotope effects normally decrease with the number of bonds or the distance from the site of isotopic substitution. Reported $\alpha-\Delta\delta_c$ are invariably positive (upfield) and range up to about 0.5 ppm per ^{2}H.

An α-deuterium shift is often more difficult to measure than a β-isotope shift, despite its greater magnitude. Particularly in the case of complete α-deuteration, a high signal-to-noise ratio is required to detect the signal of the deuterated carbon, due to the reduced NOE and multiplet appearance. The reduced intensity and splitting clearly identify the signal of a deuterated carbon atom, in some cases effectively removing it from the spectrum. For instance, α-carbons were not observed in a ^{13}C nmr study of all sixteen CD_2-isotopomers of methyl octadecanoate.[50] Thus, the α-isotope shift is probably not as generally useful a tool for signal assignment and structural identification as the other effects of α-deuteration. Nevertheless, if the magnitude of the α-deuterium shift could be correlated with structural features, the isotope shift would be an additional aid in identifying sites of isotopic substitution or in making signal assignments in complex spectra. Also, quantitative determinations of deuterium content are made possible since the signals for labelled and unlabelled carbons are not coincident due to the isotope shift.

(1) Saturated Carbons. Representative literature values of the α-deuterium isotope effect on ^{13}C chemical shifts of saturated carbons are presented in TABLE 1. The $\alpha-\Delta\delta_c$ data are divided into effects on methyl, methylene, and methine carbons. Within each category, the data are given in roughly descending order based on the magnitude of the effect per ^{2}H, although a few values are grouped together because of obvious structural relationships.

TABLE 1

One-Bond Deuterium Isotope Shifts at Saturated Carbons

Compound	$\Delta\delta_c$, ppm[a]	$\Delta\delta_c/^2H$[b]	Ref.
A) Methyl Groups			
1 PhCHDCH$_2$D	(0.4)	(0.4)	51
2 PhCHDCHD$_2$	(0.7)	(0.35)	51
3 PhCHDCD$_3$	(1.0)	(0.33)	51
4 MeO$_2$C(CH$_2$)$_{14}$CH$_2$D	0.31	0.31	52
5 MeO$_2$C(CH$_2$)$_{14}$CHD$_2$	0.60	0.30	52
6 MeO$_2$C(CH$_2$)$_{14}$CD$_3$	(0.91)	0.30	52
7 4-CH$_2$D-camphor	(0.30)	(0.30)	53
8 PhCH$_2$D	0.3	0.3	54
9 PhCHD$_2$	0.54,0.6	0.27	55,54
10 PhCD$_3$	0.86	0.29	55
11 1-CD$_3$-cyclopentanol	0.9	0.3	40
12 RR'(OH)CD$_3$	0.9	0.3	40
13 (CD$_3$)$_2$(CH$_3$)COH	0.9	0.3*	40
14 (CD$_3$)$_3$COH	1.5	0.5*	40
15 RR'(CD$_3$)C$^+$	0.9	0.3	40
16 (CD$_3$)$_3$C$^+$	1.3	0.4*	40
17 CD$_3$OD	0.86	0.29	56
18 CD$_3$OD	(0.95)[c]	(0.32)*	57
19 p-F-PhOCD$_3$	0.84	0.28	44
20 CD$_3$S(O)CD$_3$	(0.92),(1.18) 0.90,0.89	0.30*	58,57 2,56
21 HCON(CD$_3$)$_2$	(0.81)&(0.86)	(0.27)*&(0.29)*	57
22 CH$_3$(CH$_2$)$_5$CH$_2$D	0.28	0.28	59
23 CH$_3$(CH$_2$)$_{16}$CH$_2$D	(0.28)	(0.28)	60
24 CD$_3$COCD$_3$	(1.21),(0.87), 0.75,0.77	0.25*	58,57 2,56
25 PhCOCD$_3$	(0.24),0.74	0.25	61,44
26 CD$_3$NO$_2$	0.69	0.23	2
27 Me$_3$SnCD$_3$	0.700	0.23	39
28 CH$_2$DCO$_2$Me	0.2	0.2	54
29 CHD$_2$CO$_2$Me	0.4	0.2	54

TABLE 1, Continued

Compound	$\Delta\delta_C$, ppm[a]	$\Delta\delta_C/{}^2H$ [b]	Ref.
30 CH_3D	0.187	0.187	17
31 CH_2D_2	0.385	0.193	17
32 CHD_3	0.579	0.193	17
33 CD_4	0.774	0.193	17
34 CH_2DCN	0.2	0.2	54
35 CD_3CN	(0.54),(0.46), 0.45,0.44	0.15	57,61 2,56
36 CD_3Br	(0.43)	(0.14)	57
37 CD_3CO_2Na	(0.17)	(0.06)	57
38 CD_3I	(0.22),0.13	0.04	57,56

B) Methylene Groups

39 $c-C_6D_{12}$	(1.45),1.33,1.36	0.67*	58,2,56
40 4-Me-camphor-3,3-$\underline{d}_2$	(0.94)	(0.47)	53
41 $MeO_2C(CH_2)_{15}CHDMe$	0.44	0.44	60
42 $Me(CH_2)_4CHDMe$	0.4	0.4	54
43 cyclodecanone-5-$\underline{d}$	0.414	0.414	62
44 cyclodecanone-6-$\underline{d}$	0.407	0.407	62
45 $MeO_2C(CH_2)_{14}Me$,CDH at C_4, C_6, C_{14}	0.41--0.43	0.41--0.43	52
46 $Me(CH_2)_5CHDBr$	0.41	0.41	59
47 norcamphor-3,3-$\underline{d}_2$	(0.38)	(0.38)	53
48 $Me(CH_2)_5CHDOH$	0.38	0.38	59
49 cyclodecanone-2-$\underline{d}$	0.364	0.364	62
50 $CH_3CD_2CH_3$	0.72	0.36	63
51 $PhCD_2CH_3$	0.71	0.36	63
52 norcamphor-exo-3-$\underline{d}$	(0.35)	(0.35)	53
53 4-NO_2-camphor-3,3-$\underline{d}_2$	(0.63)	(0.32)	53
54 4-Cl-camphor-3,3-$\underline{d}_2$	(0.62)	(0.31)	53
55 $Me(CH_2)_{13}CDHCO_2Me$	0.30	0.30	52
56 $MeCOCHDMe$	0.3	0.3	54

TABLE 1, CONTINUED

Compound	$\Delta\delta_c$,ppm[a]	$\Delta\delta_c/^2H$[b]	Ref.
[57] Me(CH$_2$)$_5$CHDCl	0.30	0.30	59
[58] (EtOCO)$_2$CHD	(0.26)	(0.26)	61
[59] CD$_2$Cl$_2$	(0.41),(0.65),0.41	0.20	58,57,2
[60] cyclopenta-none-2,2-d$_2$	0.4	0.2	54
[61] Me(CH$_2$)$_5$CHDI	0.11	0.11	59
(C) Methine Groups			
[62] MeO$_2$C(CH$_2$)$_{15}$CDOHMe		0.52	60
[63] PhCD(Me)$_2$		0.44	55
[64] 7,7-dimethylnorbornene-1-d		(0.389)	64
[65] CDCl$_3$	(0.26),(0.21),(0.32),0.20,0.18	0.20	58,61 57,2,56
[66] CDBr$_3$		(0.11)	57

[a]Total isotope shift relative to undeuterated compound. Positive numbers are upfield shifts. Parenthetical values are from separate measurements of labelled and unlabelled compounds.
[b]Isotope shift per deuterium. Values marked by asterisk may be affected by deuteration at other sites.
[c]Relative to CH$_3$OD.

Compounds with multiple sites of deuteration, marked by asterisks in TABLE 1, have $\Delta\delta_c$ that reflect longer range effects also. For instance, the unusually high isotope shift in per-deuterocyclohexane (39) is no doubt due in part to the four β-deuterium atoms and is perhaps slightly affected by more remote deuteriums. For compounds that were examined in more than one study, the $\Delta\delta_c$ values are in excellent agreement when the $\Delta\delta_c$ values were measured directly from the peak separation in spectra of solutions containing both labelled and unlabelled material. For instance, values of 0.44 and 0.45 ppm were found for the effect of trideuteration in acetonitrile (<u>35</u>).[2,56] Values that were not obtained from solutions of paired compounds are given in parentheses in TABLE 1, and must be considered much less accurate. Unfortunately, Pohl's very extensive study[57] of isotope effects in common deuterated solvents reports values which consistently differ from, and are usually higher than, those obtained using paired compounds. Pohl's values of $\Delta\delta_c$ were from separate measurements on the pure substances, referred to TMS in a coaxial capillary.[57] Similar discrepancies may be noted in other cases, such as the unusually high $\Delta\delta_c$ reported for acetone (<u>24</u>) by Levy and Cargioli,[58] obtained from separate measurements.

TABLE 1 includes much isotope shift data that reflect multiple deuteration effects or that were determined by separate measurements. Among the remaining results, solvent, temperature, concentration, and digital resolution also vary widely, so that there is little basis for an attempt to make quantitative correlations between structure and α-$\Delta\delta_c$ values. The results in TABLE 1 should nonetheless suffice to indicate the range of effects and some possible trends.

All one-bond isotope shifts are upfield; that is, the heavier isotope shields the α-atom. α-Isotope shifts are at least approximately additive, as indicated by the precise, gas-phase results for the isotopomers of CH_4 (<u>30</u>-<u>33</u>).[17] Additivity is also seen for the sets <u>4</u>-<u>6</u>, <u>8</u>-<u>10</u>, and <u>28</u>-<u>29</u>.

The α-$\Delta\delta_c$ per ^2H atom appears to increase with increased alkyl branching at the α-carbon although directly comparable data are limited. For instance, the α-$\Delta\delta_c/^2H$ in CD_3OD (<u>18</u>) compared to CH_3OD is 0.29 ppm, increases to 0.38 for 1-heptanol-1-d (<u>48</u>) and increases further to 0.52 for a secondary alcohol (<u>62</u>). Similarly, in the series toluene (<u>10</u>), ethylbenzene (<u>51</u>), and isopropylbenzene (<u>63</u>), the corresponding α-$\Delta\delta_c/^2H$ at the benzylic carbons are 0.29, 0.36, and 0.44 ppm. There is also a monotonic increase in the series: methane (<u>33</u>), heptane-<u>1</u>-d (<u>22</u>), and propane-2,2-<u>d</u>$_2$, (<u>50</u>). Data on successive substitution of groups other than alkyl are extremely limited. The α-$\Delta\delta_c/^2H$ for methyl acetate-α-<u>d</u> (<u>29</u>) is about 0.2 ppm and the value for diethyl malonate (<u>58</u>) is 0.26 ppm. However, dichloromethane (<u>59</u>) and chloroform (<u>65</u>) exhibit the same $\Delta\delta_c/^2H$ of 0.20 ppm.

As described above, substitution of alkyl groups for hydrogen atoms at the isotopically substituted site increases the isotope shift. Using the isotope shift of 0.19 at methane[17] as the reference point, the data for methyl groups in TABLE 1 can be used to create an order of the effects of other substituents. An approximate order, starting with groups that increase the isotope shift and proceeding to those that decrease it, is as follows: alkyl, phenyl, hydroxy, phenoxy, sulfoxide > acyl, trimethyltin, nitro > ester carboxylate > hydrogen > cyano > iodide. In a study of 1-d-1-substituted heptanes, Doddrell and Burfitt found the order Br > OH > Cl > H > I.[59] The greater isotope shift at alkyl substituted carbons than at carbonyl substituted carbons is directly evident in two studies of isotopomeric compounds: cyclodecanone deuterated at C_2, C_5, or C_6 (49, 43, 44), and methyl palmitate (45 and 55) deuterated at alternate positions. Servis and Shue found the $\alpha-\Delta\delta_c$ in deuteriomethyl groups of alcohols to be insensitive to structural differences, and in fact found the same typical isotope effect in carbocations derived from the alcohols (see 11-16 in TABLE 1).[40]

The substituent effects do not exhibit any obvious relation to the electronic or steric nature of the substituents. Certainly the halogen order Br > Cl > I is difficult to rationalize in terms of either electronegativity or size of the halogen atom.[59] One type of structural relationship that has been noted in the case of other nuclei is a relation between one-bond isotope shifts and coupling constants. The effect of $^{13}C/^{12}C$ substitution on ^{19}F shifts correlates inversely with one-bond coupling constants $^1J_{CF}$,[18] and an approximate, inverse, linear relation was noted between the deuterium isotope effect on ^{31}P shifts and $^1J_{PH}$.[65] Here, some values of $\alpha-\Delta\delta_c/^2H$ are listed in TABLE 2(a) for some simple alkyl systems, along with the one-bond coupling constants, $^1J_{CH}$,[66] and C-H stretching frequencies, ν_{CH},[67] for some simple alkyl systems. The $^1J_{CH}$ and ν_{CH} are expected to provide some indication of carbon hybridization effects and C-H bond strength. No relation between $\alpha-\Delta\delta_c$ and the other two physical parameters is apparent, as was noted previously by Gold in regard to a correlation with coupling constants.[2]

While no obvious structural relationship exists for $\alpha-\Delta\delta_c$ in variously substituted alkyl systems, a clear trend toward increased isotope effects with increased ring size is evident in the cycloalkanes. Günther recently measured at high field (9.4 T, 100.6 MHz ^{13}C) the isotope shifts in monodeuterio cycloalkanes

TABLE 2
One-bond Isotope Effects, Coupling Constants, and
Stretching Frequencies

Compound	$\alpha - \Delta\delta_C / {}^2H$ (ppm)	${}^1J_{CH}$ (Hz)	ν_{CH} (cm^{-1})
(a) Alkyl Systems			
$PhCH_3$	0.29	126	2953
CH_3OH	0.29	141	2920, 2979
$CH_3S(O)CH_3$	0.30	137	
CH_3COCH_3	0.25	127	3000, 2946
$PhCOCH_3$	0.25	126	
$(CH_3)_4Sn$	0.23	128	2960
CH_4	0.19	125	2992
CH_3CN	0.15	136	2979
CH_3I	0.04	151	3029
CH_2Cl_2	0.20	178	3025
$CHCl_3$	0.20	209	3034
(b) Cycloalkanes[68]			
$c-C_3H_6$	0.309	160.3	
$c-C_4H_8$	0.363	133.6	
$c-C_5H_{10}$	0.374	128.5	
$c-C_6H_{12}$	0.418	125.1	
$c-C_7H_{14}$	0.412	123.6	

from cyclopropane to cycloheptane.[68] These results are given in
TABLE 2(b), along with one-bond coupling constants.[68] There is a
rough inverse relation between the two parameters. Since $^1J_{CH}$
are known to be related to the s character in the carbon hybrid
orbital forming the C-H bond, the $\alpha - \Delta\delta_C$ appear to increase with
increasing p character and should increase with C-H bond
lengths.[68]

(2) <u>Unsaturated</u> <u>Carbons.</u> The only report of a direct determination of an α-deuterium effect on a ^{13}C shift of an alkene carbon is that of Doddrell and Burfitt for hept-1-ene-1-<u>d</u>, which was a mixture of the E- and Z- isomers.[59] The $\alpha-\Delta\delta_c$ for hept-1-ene-1-<u>d</u> is 0.21 ppm, somewhat lower than the 0.28 ppm for the saturated heptane-1-<u>d</u>.[59] Friesen and Wasylishen recently reported an $\alpha-\Delta\delta_c$ of 0.16 for hydrogen cyanide.[69] Paquette, Mason <u>et</u> <u>al</u>., gave ^{13}C shifts for apobornene-2-<u>d</u>, (<u>67</u>), which indicate an upfield shift of 0.13 ppm for the deuterated alkene position relative to the β-alkene carbon, but an absolute $\alpha-\Delta\delta_c$ was not determined by comparison with the unlabelled compound.[64] Saunders found $\alpha-\Delta\delta_c$ values of 0.25 and 0.29 ppm for the labelled positions in the cyclohexenyl (<u>68</u>) and cyclopentenyl (<u>69</u>) cations.[70]

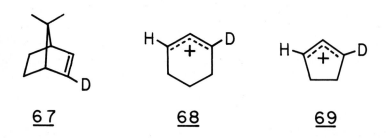

<u>67</u>   <u>68</u>   <u>69</u>

Isotope shift data for benzenoid aromatic hydrocarbons are sufficiently extensive to suggest a typical value of 0.30±0.02 ppm for $\alpha-\Delta\delta_c$. Bell <u>et</u> <u>al</u>. found the α-isotope shift in benzene-1-<u>d</u> to be 0.289±0.003 ppm, and also found β and γ shifts of 0.110 and 0.011 ppm, respectively.[71] A more recent study, which determined shifts relative to C_4 rather than through comparison with unlabelled benzene, found essentially the same results for benzene-1-<u>d</u>.[72] The isotope shift of perdeuterated benzene has been determined several times.[2,43,57,58] In the most accurate of these, Gold <u>et</u> <u>al</u>. observed a $\Delta\delta_c$ of 0.530±0.005, predictable on the basis of additivity of α and remote effects from the benzene-1-<u>d</u> results (predicted: 0.289 + 2 x 0.110 + 2 x 0.011 = 0.531).[2] The $\alpha-\Delta\delta_c$ is 0.31 for naphthalene-1-<u>d</u>, 0.28 for naphthalene-2-<u>d</u>, 0.31 for phenanthrene-9-<u>d</u>, and 0.27 for benzo[c]phenanthrene-2-<u>d</u>.[13] Effects of similar magnitude were seen in 1-substituted 4-<u>d</u>-naphthalenes.[73] Anthracene-9,10-<u>d</u>$_2$ shows a higher $\Delta\delta_c$ of 0.38 ppm,[13] which may be indicative of a long-range effect of the second deuterium atom. Buchanan and Ozubko found isotope shifts of about 0.31 ppm at three different deuterium-labelled positions in benzo[a]pyrene.[74]

Bell et al. examined deuterium isotope effects in substituted benzenes.[71] Ortho-, meta-, and para-monodeuterated anisole, toluene, and chlorobenzene were studied, and also benzonitrile-2-d and benzotrifluoride-4-d. The substituent influence on α-isotope shifts correlated with the electronic effects of the substituents as described by the Hammett substituent constants σ_o, σ_m, and σ_p, with the larger isotope shifts associated with the more electron-releasing substituents. The range of isotope shifts was small in the m- and p-substituted series (α-$\Delta\delta_c$ from 0.276 to 0.295 ppm), but was distinctly larger in the ortho series (0.243 to 0.338 ppm).[71] An earlier study of the isomeric deuteriobenzonitriles did not directly compare labelled and unlabelled compounds in the same solution.[75] The lack of a significant steric effect on the isotope shift at the ortho position is suggested by comparing the α-$\Delta\delta_c$ of 0.31 ppm in t-butylbenzene-2-d, found by Maciel et al.,[55] with the value of 0.321 ppm for toluene-2-d.[71] Isotope shifts in a few perdeuterated benzenes have been reported but were not measured directly in binary solutions.[57]

α-Isotope shifts have been measured in only a few nonbenze-noid aromatics. The ^{13}C signal in perdeuterated ferrocene is shifted 0.40 ppm upfield,[76] which is a shift comparable to those in benzenoid aromatics. However, the same study found isotope shifts an order of magnitude higher, and both upfield and downfield isotope shifts, in paramagnetic metallocenes such as $(C_2H_5)_2Fe^+$ and $(C_2H_5)_2Co$. The ^{13}C chemical shifts are widely shifted from the ordinary range for diamagnetic materials and the $\Delta\delta_c$ are less than 1.5% of these paramagnetic shifts. The $\Delta\delta_c$ apparently arise from an isotope effect on the interaction of the ring carbons with the unpaired electrons.[76]

Isotope shifts in another nonbenzenoid aromatic, the tropylium-1-d ion (70), were measured with high precision, but unfortunately were not referenced to unlabelled tropylium ion.[72] Günther et al. assumed that the three-bond isotope effect was zero, resulting in values shown below.[72] This assumption was made in order to keep all isotope shifts positive, but Saunders has demonstrated negative three-bond isotope shifts of -0.167 and -0.163 ppm in the related cyclohexenyl (68) and cyclopentenyl (69)

$$\alpha-\Delta\delta_C \qquad 0.408 \qquad 0.373$$

$$\beta-\Delta\delta_C \qquad 0.133 \qquad 0.098$$

$$\gamma-\Delta\delta_C \qquad 0^* \qquad -0.035 \qquad * = \text{assumed}$$
$$\text{value}$$

$$\delta-\Delta\delta_C \qquad 0.035 \qquad 0^*$$

70

cations.[14]   Clearly, all four isotope shifts could be nonzero. Possible alternative values of isotope shifts are given above, based on the assumption of a negligible four-bond isotope effect.

(3)   Applications of α-Deuterium Isotope Effect.   Probably the most important application of the α-deuterium isotope effect on ^{13}C chemical shifts arises from the need to recognize that the ^{13}C resonances in perdeuterated solvents are shifted by as much as one ppm from the undeuterated substances.   Deuterated solvents are commonly used to provide a ^2H nmr signal for internal field/-frequency lock and to reduce the size of interfering solvent signals in the ^{13}C spectrum.   If solvent signals are to be used as secondary references for chemical shifts, it is necessary to use the isotopically shifted δ_C values.

Chemical shifts for common deuterated solvents are given in TABLE 3, from four sources.   Pohl et al. reported shifts relative to TMS in a coaxial capillary.[57]   Some of the most common solvents were listed in a study by Levy and Cargioli.[58]   In a text on ^{13}C nmr spectroscopy, Wehrli and Wirthlin listed chemical shifts for deuterated solvents in a correlation chart, but did not refer to the original source.[19]   With the exception of dioxane, the values of Wehrli and Wirthlin match those in a data sheet widely distributed by Merck, Sharp, and Dohme, Inc., but not yet published in the open literature.   The Merck ^{13}C data were determined at 25 MHz on solutions containing 5% TMS.

Deuterium labelling is a common technique that simplifies ^{13}C spectral interpretation by weakening the signal at the isotopically labelled positions.   For instance, Roberts et al. assigned α-carbon resonances in keto steroids after proton-deuteron exchange at the relatively acidic sites α to the carbonyl groups.[1]

Despite the common practice of deuterium labelling, the literature does not appear to contain any clearcut examples in which an intrinsic one-bond isotope effect of deuterium substitution was a key to structural identification or ^{13}C signal assignment. An instance which approaches closely to such an application is a study by Bovey et al. of poly(vinyl chloride) that had been reduced by LiAlD$_4$.[77] The triplet for methylene carbons where a chlorine atom had been replaced by a deuterium were more easily identified in the complex spectrum because they were shifted relative to unlabelled methylenes.[77] Perhaps the most significant

TABLE 3
^{13}C Chemical Shifts of Perdeuterated Solvents

Compound	δ_C (ppm from TMS)			
	Pohl*[57]	Levy[58]	Wehrli[19]	Merck
Acetic Acid	18.25		20.0	20.0
	176.6		178.4	178.4
Cyclohexane	25.14	26.06		26.4
Acetone	28.05	29.22	29.8	29.8
	204.35		206	206.0
Dimethylsulfoxide	39.60	39.56	39.5	39.5
Methanol	47.07		49.0	49.0
Dichloromethane	53.08	53.61	53.8	53.8
Dioxane	65.38		67.4	66.5
Tetrahydrofuran	24.06			67.4
	66.03			25.3
Chloroform	76.93	76.91	77.0	77.0
Pyridine	122.55		123.5	123.5
	134.33		135.5	135.5
	148.68		149.9	149.9
Benzene	127.88	127.96	128.0	128.0

*Measured relative to external (capillary) TMS.

use in structural studies of the intrinsic α-isotope effect on ^{13}C shifts is to provide a reference standard for isotope effects in non-equilibrating systems. Isotope effects which differ significantly from the ordinary are indicative of an isotopically perturbed equilibrium.[14,70,78]

The α-isotope shift has been used to advantage in quantitative studies of deuteration.[3,54,75,79,80] Since the $\alpha\text{-}\Delta\delta_C$ per deuterium is usually less than 0.5 ppm and $^1J_{C-D} \geq 18$ Hz, the ^{13}C-^{2}H induced triplet and the residual ^{13}C-^{1}H signal will not overlap at the operating frequencies of most ^{13}C instruments (15-25 Hz/ppm). Separate measurements can then be made of signal intensities for deuterated and undeuterated carbon. Stothers demonstrated that direct comparison of the integrated intensities of partially deuterated methylenes or methyls to the intensities of the corresponding unlabelled carbons provides a direct measure of deuterium content.[3,79,80] Direct comparison is possible because the nuclear Overhauser enhancement from proton decoupling is the same at a particular site, independent of the number of ^{1}H atoms attached, except in the case of a fully deuterated carbon. At a fully deuterated carbon, the signal intensity generally will not be useful because the NOE can be considerably reduced and saturation effects may be important because of a longer T_1. In such cases, deuterium content can be determined by calibrating the intensity of the residual ^{13}C-^{1}H signal to signal intensity in an unlabelled sample, or to a suitable reference peak in the labelled material, or to a peak of an added calibration substance.[81] In the case where only a small amount of perprotio compound is present, Stephenson found it useful to record the spectrum before and after spiking the sample with a known amount of perprotio material.[54] The spiking technique would be useful regardless of the completeness of deuteration of the labelled material.

These techniques of site-specific, quantitative deuterium analysis by ^{13}C nmr have been used in the analysis of deuteriobenzonitriles,[75] of reduction products in the reaction of halides with Zn-Cu and D_2O,[54] of exchange processes in ketones and aromatics,[80,81] and of cationic rearrangements.[79] Similarly, hydrogen exchange during the biosynthesis of fatty acids was determined quantitatively by integration of isotopically shifted ^{13}C resonances in methyl palmitate.[52] In this biosynthesis study, the analysis was simplified by simultaneous ^{1}H and ^{2}H decoupling to create all singlet resonances.[52]

(B) Isotope Shift Induced by Other Nuclei

(1) ^{18}O-Induced Shifts. Several reports have appeared recently concerning the ^{16}O/^{18}O isotope effect on ^{13}C chemical shifts.[20,82-86] The effect is invariably an upfield shift for the isotopomer containing the heavier isotope. Even though $\alpha-\Delta\delta_c(^{18}$O) is smaller than $\alpha-\Delta\delta_c(^2$H), the $\alpha-^{18}$O isotope shift is potentially of much greater value than the $\alpha-^2$H isotope shift because the other effects associated with deuterium labelling do not occur with ^{18}O labelling. Both ^{16}O and ^{18}O are nmr inactive (I = 0) nuclei, which means there will be no change in coupling constants or peak multiplicity and no alteration of signal intensity due to changes in NOE or relaxation times. The isotope shift is essentially the only nmr means of detecting the isotopic substitution. The ^{18}O isotope shift is particularly promising in regard to mechanistic studies where it is desired to follow the scrambling or exchange of an ^{18}O label. The analogous ^{18}O-induced shift in ^{31}P nmr also appears very promising as a probe for following the reactions of phosphates.[20]

Risley and Van Etten,[20,82,86] and Vederas[85] have explored the dependence of $\alpha-\Delta\delta_c(^{18}$O) on structure in organic compounds. The isotope shift is 0.020 to 0.026 ppm in primary and secondary alcohols, and 0.030 to 0.035 in tertiary alcohols. The trend among closely related compounds is clearly $3° > 2° > 1°$, with the larger difference occurring between tertiary and secondary. For instance, in the series t-butyl alcohol, isopropyl alcohol, and n-butanol, the respective isotope shifts are 0.035, 0.023, and 0.020 ppm.[20,86] This is the same order seen for deuterium isotope effects, discussed in SECTION IIA1. Phenyl substitution at C_1 results in slightly smaller shifts: 0.019 in benzyl alcohol[85] and 0.025 in triphenylcarbinol.[86] Steric crowding is apparently not the source of the larger isotope shifts in tertiary alcohols, as tri-t-butylcarbinol actually has a smaller isotope shift than t-butyl alcohol.[86] Phenol has a smaller $\alpha-\Delta\delta_c(^{18}$O) of 0.016 ppm.[20] The smaller isotope shift associated with an oxygen singly bound to an sp^2 carbon is also observed in ethers and esters, as shown below for phenyl ethyl ether (71),[20] phenyl vinyl ether (72),[20] and n-propyl benzoate (73).[85]

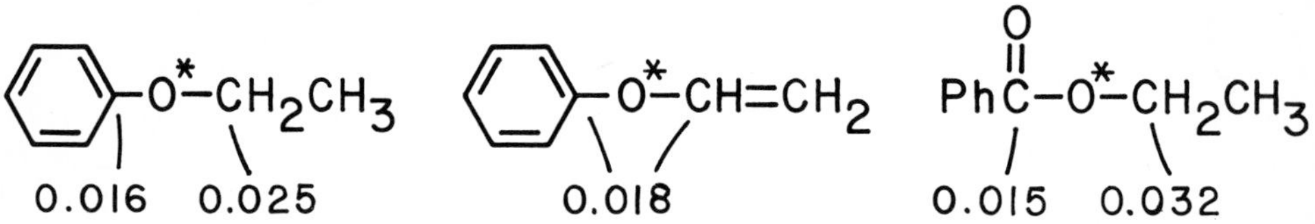

0.016 0.025   0.018   0.015 0.032

71   **72**   **73**

The ^{18}O-induced isotope shift has been examined for carbonyl compounds also. The largest effects are in aldehydes and ketones, with values ranging from 0.042 to 0.053 ppm.[20,85,86] Vederas found that $\alpha\text{-}\Delta\delta_c(^{18}O)$ decreases in the order ketones $\geq$ aldehydes $\geq$ esters $\geq$ amides.[85] This order parallels the order of decreasing C-O double bond character. The carbonyl-^{18}O effect in esters is about 0.036 ppm while the alkoxy-^{18}O effect is about 0.015 ppm. Since the two effects are approximately additive the isotope effect provides a convenient means of determining the location and extent of [18]O labelling in esters.[20,85,86] Risley and Van Etten demonstrated the use of ^{13}C peak areas of isotopically shifted peaks as a quantitative measure of [18]O content in the first report of the [18]O isotope effect on ^{13}C nmr shifts.[82]

The ^{18}O-induced isotope shift of ^{13}C resonances in metal carbonyls provides a stereochemically specific, quantitative method for tracing labelled carbon monoxide ligands. The $\alpha\text{-}\Delta\delta_c(^{18}O)$ in group 6B metal carbonyls is 0.040 to 0.045 ppm.[83,84] Darensbourg used the isotope shift to follow inter- and intra-molecular processes in some tungsten carbonyls.[83,84]

(2) <u>Shifts Induced by ^{13}C, ^{15}N, [34]S, [37]Cl</u>. Isotope shifts induced in ^{13}C resonances by various other nuclei have been observed, but are not likely to be as useful as the [18]O shift due to greater difficulty in synthesizing the labelled compounds, or a very small magnitude of the isotope effect, or the occurrence of more obvious changes upon substitution, such as a change in spin-spin coupling. As usual, the one-bond effects are all upfield shifts.

The $^{12}C/^{13}C$ effect is 0.018 ppm in ferrocene,[87] 0.03 ppm in thiophene for both C_3 acting on C_2 and C_2 on C_3,[88] and 0.01 ppm in 3-pentanol for both C_1, C_2 interactions.[88] In the four-membered rings oxetane, thietane, and cyclobutane, the typical $\alpha-\Delta\delta_c(^{13}C)$ is 0.012 ppm, but in cyclobutanone an unusually large value of 0.027 ppm was found for the effect of the C_1 carbonyl carbon on C_2, and an unusually small value of 0.002 ppm for C_2 acting on C_1.[89] Spin-spin coupling between ^{13}C nuclei is a more important phenomenon, reflecting hybridization and other structural features.[66,88,90]

The $^{14}N/^{15}N$ effect on the ^{13}C shift in the cyanide ion and some metal cyano complexes is 0.02 to 0.04 ppm,[91] which is slightly smaller than the $^{16}O/^{18}O$ effect reported for metal carbonyls. The observation of a small $^{35}Cl/^{37}Cl$ effect has been claimed for some polychlorinated hydrocarbons, amounting to about 0.004 ppm per ^{37}Cl(0.1 Hz at 25 MHz).[92] Since the $^{35}Cl:^{37}Cl$ ratio of natural abundance is 3:1, the isotope effect, as well as quadrupole-induced relaxation by ^{35}Cl and ^{37}Cl, may contribute to broadening of the solvent signals in ^{13}C spectra run in $CDCl_3$ or CCl_4. A small (0.009 ppm) isotope effect of $^{32}S/^{34}S$, at the 4.2% natural abundance of ^{34}S, has been observed in carbon disulfide, another common ^{13}C nmr solvent and reference substance.[93]

III. TWO- AND THREE-BOND ISOTOPE EFFECTS ON ^{13}C SHIFTS

(A) Effects of Deuterium Bonded to Carbon

The two- and three-bond effects (β- and γ-effects) of deuterium substitution on ^{13}C chemical shifts are smaller than the α-effect, but inherently more valuable. The β-effect is typically one-third to one-quarter the magnitude of the α-deuterium shift, i.e., roughly 0.1 ppm per deuterium. A $\beta-\Delta\delta_c$ of this magnitude can be reliably detected in a ^1H-decoupled ^{13}C spectrum by modern instruments, even at low field strength, provided that the lines are not broadened by poor tuning or other conditions. There are only a few reports of γ-isotope shifts, which are clearly smaller yet, perhaps again by a typical factor of about one-third to one-quarter. The γ-shift may require a high field instrument for routine detection. Both β- and $\gamma-\Delta\delta_c$ should be determined from binary solutions containing both the labelled and unlabelled compounds, because the small magnitude of the effects is comparable to normal variation in chemical shifts in separate determinations.

The β- and γ-isotope shifts are potentially valuable phenomena for aiding signal assignment in ^{13}C spectra. Isotopic labelling can thereby distinguish not only the directly labelled position but nearby carbon atoms also. The β-shift is the most convenient way to identify the nmr signal of a carbon next to a deuterated carbon, because it is usually the only obvious change produced in a ^{1}H-decoupled spectrum by β-deuteration. Signal intensity will not be altered by β-deuteration, unless the carbon in question is itself not bonded to protons. In that case, the NOE and spin-spin relaxation is affected by β-deuteration. Also, two-bond coupling constants to hydrogen, J_{CCH}, are usually small, such that the coupling to deuterium, J_{CCD}, may be so small as to not even noticeably broaden the carbon resonance. However, three-bond couplings to hydrogen, J_{CCCH}, are usually larger than two-bond couplings, so that a γ-carbon may be distinguished by a small J_{CCCD} or broadening, as well as the small isotope shift. The β-isotope shift is actually easier to measure than the α-isotope shift, at least if the α-carbon is completely deuterated, because the signal of the deuterated α-carbon is reduced in intensity and split due to one-bond coupling, J_{CD}. The β- and γ-$Δδ_c$ are usually upfield shifts, but shifts at carbonyl carbons and carbocation centers are important exceptions.

(1)   <u>Isotope Shifts at Saturated Carbons.</u>   Accurate β-$Δδ_c$ values for the effect of deuterium substitution in saturated chains are scarce, but the available data indicate a range of 0.06 to 0.12 ppm per ^{2}H. Doddrell and Burfitt found C_2 shifted by 0.11 ppm in both 1-bromo- and 1-chloroheptane-1-$\underline{d}$.[59] β-Isotope shifts were not resolved in heptane-1-$\underline{d}$, 1-iodoheptane-1-$\underline{d}$, and 1-heptanol-1-$\underline{d}$. Maciel <u>et al</u>. reported a β-$Δδ_c$ of 0.10 ppm for the methyl carbons in 2-phenylpropane-2-$\underline{d}$, but only a shift of 0.15 ppm for the effect of double labelling in phenylethane-1,1-$\underline{d}_2$.[55] Bovey investigated branching in poly(vinyl chloride) by reduction to a hydrocarbon with LiAlD$_4$, followed with characterization by ^{13}C nmr.[77] The reduced polymer exhibits a β-$Δδ_c$ of 0.21 ppm at a methylene flanked by two CHD units, in the primary chain of the

polymer. However, near the branching points, the β-isotope effect is lower, as shown below.

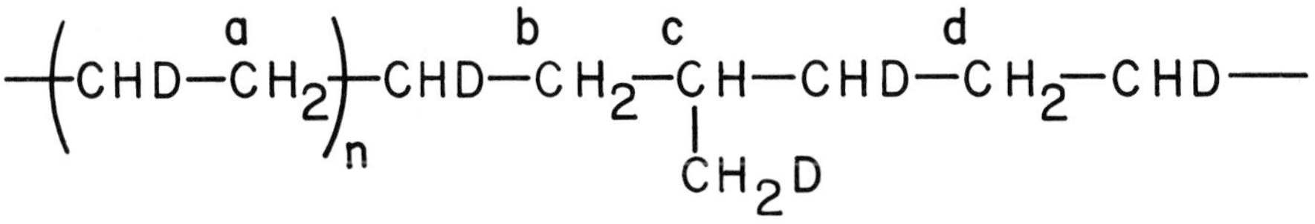

$$\left(\!\!\!\overset{a}{\underset{n}{CHD-CH_2}}\!\!\!\right)\!\!\overset{b}{CHD-CH_2}\!\overset{c}{CH-CHD}\!\overset{d}{-CH_2-CHD-}$$
$$\underset{CH_2D}{|}$$

Position	$\Delta\delta_c$	Number of ^2H neighbors
a	0.21	two
b	0.08	one
c	0.13	two
d	0.11	two

Tulloch thoroughly studied isotope effects on ^{13}C shifts in methyl octadecanoate, $CH_3O_2C(CH_2)_{16}CH_3$, and related structures.[50,60] In the first communication,[60] it is not stated whether the isotope shifts were measured directly, using binary mixtures, but the shifts are about the expected magnitude. Octadecane-1-d has a $\beta\text{-}\Delta\delta_c$ of 0.08 ppm at C_2 and a $\gamma\text{-}\Delta\delta_c$ of 0.02 ppm at C_3.[60] The $\Delta\delta_c/^2H$ appear to be larger when the C_{18} methyl is triply deuterated in methyl octadecanoate-18-$\underline{d}_3$: the $\beta\text{-}\Delta\delta_c/^2H$ at C_{17} is 0.12 ppm and the $\gamma\text{-}\Delta\delta_c/^2H$ is 0.05 ppm. Monodeuteration of the ester at C_{17} produces $\beta\text{-}\Delta\delta_c$ of 0.12 and 0.16 ppm at C_{16} and C_{18}, and a $\gamma\text{-}\Delta\delta_c$ at C_{15} of 0.02 ppm. Dideuteration at C_{17} gives values of $\beta\text{-}\Delta\delta_c$ equal to 0.24 and 0.28 ppm at C_{16} and C_{18}, and a $\gamma\text{-}\Delta\delta_c$ of 0.08 ppm. The shifts induced by successive deuteration are not all strictly additive. Similarly, the reported isotope shift of 0.48 ppm at C_{17} in methyl octadecanoate-16-$\underline{d}_2$-18-$\underline{d}_3$[60] does not match the 0.57 ppm expected on the basis of additivity from $\beta\text{-}\Delta\delta_c$ values of 0.36 ppm in the 18-$\underline{d}_3$ ester[60] and 0.21 ppm in the 16-$\underline{d}_2$ ester.[50] Because the additivity question is important in both a practical and theoretical sense, it is unfortunate that it is not clear in this study whether the isotope shifts are accurate values, directly measured from binary mixtures.

In the second paper, Tulloch reported isotope shifts for all sixteen isotopomers of methyl octadecanoate with a double deuterium label in a methylene group.[50] For C_5 through C_{17}, the average $\beta-\Delta\delta_C$ was 0.20 ppm and the average $\gamma-\Delta\delta_C$ was 0.05 ppm for the effect of two deuterium atoms. For instance, the isotope shifts in methyl octadecanoate-5,5-$\underline{d}_2$ are shown below.

$$CH_3(CH_2)_{10}-CH_2-CH_2-CD_2-CH_2-CH_2-CH_2\,CO_2\,CH_3$$

$$0.05 \quad 0.20 \qquad 0.19 \quad 0.04$$

In the vicinity of the carbonyl carbon, the effects were reduced. C_4 deuteration shifted C_3 by only 0.15 ppm, C_3 deuteration shifted C_2 by 0.18 ppm, C_2 deuteration shifted C_3 by 0.14 ppm, and no isotope effect was detected at the carbonyl carbon. Other effects of deuteration were to broaden $\beta-$ and γ-carbon resonances due to coupling with ^2H and to reduce the signal intensity of the α-carbon to the extent that it was not observed.

In an interesting demonstration of the application of isotope shifts to signal assignment in ^{13}C spectra, Tulloch[50] used the isotope effects established from the isotopomers of specifically dideuterated methyl octadecanoate to assign signals in seven specifically deuterated oxo derivatives of methyl octadecanoate. From these assignments, the substituent effect of the oxo group was established and in turn used to assign chemical shifts in all sixteen isomeric oxooctadecanoates.

Another elegant application of β-isotope shifts is to signal assignment in ^{13}C nmr spectra of carbohydrates.[94-96] Gorin prepared and measured isotope shifts for two specifically deuterated derivatives of most of the frequently encountered mono-saccharides and some of the derived methyl glycosides. The α-carbon isotope effects and the β-isotope shifts permitted unambiguous assignments of each signal in both α- and β-anomers of each sugar, and resulted in several changes in assignments from previous studies. The $\beta-\Delta\delta_C$ for monodeuteration were 0.04 to

0.09 ppm, and 0.07 to 0.13 ppm for dideuteration.[95] An unusually high value of 0.17 ppm was found for C_3 in β-D-arabinopyranose-4,5,5-$\underline{d}_3$, while the isomeric α-D-arabinose displayed a normal value of 0.10 ppm.[95] In all other cases, the configurational difference was $\leq$ 0.02 ppm. No variation was noted on going from an axial to an equatorial ^{2}H nucleus.[94] However, β-isotope shifts at C_1 are considerably reduced, at $\leq$0.02 ppm, than β-$\Delta\delta_c$ at other positions. This was ascribed to the effect of having two electro-negative oxygen atoms bonded to C_1, because it was also noted that the typical range of β-$\Delta\delta_c$ in sugars is lower than in compounds with less electronegative substituents attached.[94] In another carbohydrate study,[96] protons on carbons with hydroxyl groups attached were exchanged with D_2O in the presence of Raney nickel, and the β-$\Delta\delta_c$ were found to be about 0.06 ppm.

A γ-$\Delta\delta_c$ of 0.088 ppm was determined for the effect of CD_3 on the CH_3 resonances in tetramethyltin.[39] This γ-effect is comparable to those observed in hydrocarbon chains. However, the β-$\Delta\delta_c$ on the ^{119}Sn resonance in $Sn(CH_3)_3CD_3$ is considerably larger, at 0.80 ppm, than β-effects on ^{13}C resonances. The larger effect was attributed to the larger size of the paramagnetic shielding tensor in ^{119}Sn.[39]

Günther's recent determination of ^{2}H isotope shifts in monodeuterated cycloalkanes at high field strength (9.4T, 100.6 MHz ^{13}C) is the most precise study of β- and γ-effects to date.[68] The results for cycloalkanes through cycloheptane are as follows: (cycloalkane, β-$\Delta\delta_c$, γ-$\Delta\delta_c$) C_3H_6, 0.064; C_4H_8, 0.147, 0.027; C_5H_{10}, 0.103, -0.012; C_6H_{12}, 0.104, 0.025; and C_7H_{14}, 0.110, 0.027. In addition, a downfield, four-bond $\Delta\delta_c$ of -0.014 was observed in cycloheptane, although no four-bond effect was detected for cyclohexane. The most interesting feature is the occurrence of downfield shifts at C_3 in cyclopentane-1-$\underline{d}$ and C_4 in cycloheptane-1-$\underline{d}$. The interpretation that such downfield shifts are probably intrinsic, and not equilibrium effects due to isotopic perturbation of conformational equilibria, was supported by finding a downfield $\Delta\delta_c$ of -0.037 at C_4 in conformationally rigid norbornane-1-$\underline{d}$. Günther's ^{13}C nmr spectra for cyclopentane-1-$\underline{d}$ and cyclohexane-1-$\underline{d}$ are reproduced below in FIGURE 2.

The cycloalkane data of Günther also point out that while the isotope effect is attenuated with distance or the number of intervening bonds, the relationship is not a simple one.[68] The one-bond effect is largest in cyclohexane and cycloheptane (0.418 and 0.412 ppm, respectively) and smallest in cyclopropane (0.309 ppm). The two-bond effect is also smallest in cyclopropane, but cyclobutane exhibits the largest two-bond effect. The downfield $\gamma-\Delta\delta_C$ in cyclopentane is also inexplicable by a simple scheme relying on attenuation with distance or intervening bonds.

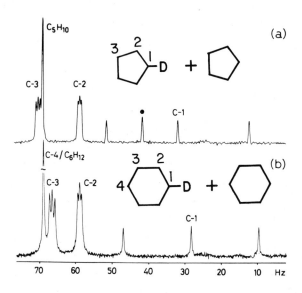

FIGURE 2. 100.61-MHz ^{13}C NMR spectra of (a) cyclopentane/cyclopentane-1-d̲ and (b) cyclohexane-1-d̲ with ^{1}H broadband decoupling; signal marked with ● is due to unknown impurity. Reproduced by permission of the American Chemical Society.[68]

(2)   Isotope Shifts at Unsaturated Carbons

(a) **Alkenes and Alkynes.** Isotope shift data for alkenes and alkynes are virtually nonexistent. Doddrell and Burfitt were not able to resolve a β-isotope shift for a mixture of E- and Z-1-heptene-1-d̲, but found the exceptionally large $\beta-\Delta\delta_C$ of 0.50 ppm in 1-heptyne-1-d̲.[59]   The $\beta-\Delta\delta_C$ in 1-heptyne-1-d̲ is more than twice as large as the $\alpha-\Delta\delta_C$ of 0.22 ppm. Apparently, there are no further literature data to confirm this exceptional result for an alkyne. The only isotope shift data for alkenes is a report of differences in chemical shifts in 1- and 2-deuterated 7,7-dimethylnorbornenes for positions which are equivalent in the unlabelled material.[64] Absolute isotope shifts were not determined by direct comparison with unlabelled material.

(b) Aromatics. Isotope shift data are more extensive for benzenoid aromatic hydrocarbons, as noted in SECTION IIA2 for α-isotope shifts. The β-isotope shift is one of three effects of deuteration that have practical importance in ^{13}C nmr spectra of arenes.[73,81] In specifically deuterated arenes, the α-carbon resonance is easily identified by the reduced signal intensity and triplet appearance, a β-carbon can be identified by the β-isotope shift, and a γ-carbon gives a broadened signal or triplet due to three-bond coupling to ^2H. The β- and γ-effects in benzene are typical. Bell, Chan, and Sayer found the ortho carbons in benzene-1-d to be shifted upfield by 0.110 ppm.[71] The coupling of the ortho carbons to deuterium, J_{CCD}, is expected to be only 0.15 Hz, based upon the J_{CCH} = 1.0 Hz in benzene.[97] The meta carbons in benzene-1-d show a much smaller γ-$\Delta\delta_c$ of 0.011 ppm, but have a larger coupling, J_{CCCD} = 1.14 Hz, corresponding to J_{CCCH} = 7.4 Hz in benzene.[71,97]

Bell et al. found only small variations in β-$\Delta\delta_c$ from the benzene value of 0.110 ppm for a series of monodeuterated, substituted benzenes.[71] With the exception of C_1 in anisole-2-d, the values ranged from 0.086 ppm at C_1 in toluene-2-d to 0.123 ppm at $C_{3,5}$ in toluene-4-d, with little apparent relationship to the electronic character of the substituent. The β-$\Delta\delta_c$ at C_1 in anisole-2-d was only 0.037 ppm. Bell concluded that the β-$\Delta\delta_c$ were dominated by the effects of the substituent on the vibrational modes of the ring, rather than electronic effects.

In a study of isotope shifts in polycyclic aromatic hydrocarbons, Martin, Morian, and Defay found β-$\Delta\delta_c$ varying from 0.05 to 0.12 ppm.[13] In arenes where two different β-$\Delta\delta_c$ occur for nonequivalent carbons ortho to the deuterated site, the larger β-$\Delta\delta_c$ is associated with the larger π-bond order to the deuterated carbon.[13] For instance, the β-$\Delta\delta_c$ for phenanthrene-9-d are 0.12 and 0.07 ppm at C_{10} and C_{8a}, respectively, while the corresponding π-bond orders (p_{rs}) are $P_{9,10}$ = 0.775 and $P_{8a,9}$ = 0.506. Martin et al. did not observe a γ-$\Delta\delta_c$ for meta carbons but did find a γ-$\Delta\delta_c$ for peri carbons of about 0.04 ppm. Both meta and peri carbons were broadened by three-bond couplings to deuterium. Thus, the effects of deuterium labelling in benzenoid hydrocarbons can be used to assign four types of ^{13}C signals: C_α-D, C_β-C-D ortho, C_γ-C-C-D meta and C_γ-C-C-D peri.[13]

α: $\alpha-\Delta\delta_c$, J_{CD}, weak signal

β: $\beta-\Delta\delta_c$

meta γ: J_{CCCD}

peri γ: $\gamma-\Delta\delta_c$, J_{CCCD}

Kitching, Bullpitt, Doddrell, and Adcock demonstrated a similar high utility of deuterium isotope effects in making ^{13}C assignments in 1- and 2-naphthyl compounds.[73] The $\beta-\Delta\delta_c$ of about 0.1 ppm and broadening due to J_{CCCD} were keys to several assignments. This typical $\beta-\Delta\delta_c$ was also used for signal assignments in phenyl substituted anthracenes[98] and benzo[a]pyrene.[74] Stothers et al. reassigned signals for 1,8-dimethylnaphthalene, based on $\beta-\Delta\delta_c$ and the other effects of deuteration.[81]

Substitution of N-D for N-H in pyrrole produces slightly larger upfield shifts than is typical of the benzenoid hydrocarbons. The $\beta-\Delta\delta_c$ in pyrrole-N-d is 0.177 ppm and the $\gamma-\Delta\delta_c$ is 0.034 ppm.[99] As discussed in SECTION IIA2, monodeuteration of the tropylium ion produces isotope shifts comparable to those in benzene,[72] although absolute values were not determined. Deuteration of a side chain position may induce smaller isotope shifts in the ring, per deuterium, judging from the only available example of toluene-$\alpha,\alpha,\alpha-d_3$, with β- and $\gamma-\Delta\delta_c$ of 0.10 and 0.01 ppm, respectively.[55]

(c) Ketones and Other Polar Groups. In contrast to isotope shifts for most structural types in which substitution of a heavier nucleus causes an upfield shift, downfield shifts have been observed at carbonyl carbons and in carbocations. Maciel and Ellis reported the first such result, a downfield $\beta-\Delta\delta_c$ of -0.28 ppm at the carbonyl carbon of acetone-d_6.[30] Pohl also reported the downfield shift in acetone.[57] Recently, Servis and Shue found downfield β-shifts in some stable carbocations, such as the t-butyl cation, $(CD_3)_3C^+$, for which $\beta-\Delta\delta_c = -1.4$ ppm.[40] The origin of the downfield shift for both the carbocations and the polar carbonyl group is probably a hyperconjugative interaction of the electron deficient carbon with the C-H(D) bonds.[40] The C-H bond acts as a better hyperconjugative electron donor than a C-D bond, so deuterium substitution deshields the carbonyl or cation center by lowering the electron density. The hyperconjugative interaction and results for carbocations are discussed in SECTIONS IB and IVB.

A few other ketones have also been examined, with some interesting variation in results depending on structure. Morris and Murray examined some substituted camphors, including 4-methyl-camphor, 74, and norcamphor, 75, for the effect of deuteration adjacent to the carbonyl,[53] and Wehrli et al. studied three mono-deuterio isotopomers of cyclodecanone, 76.[62] Partial results were reported for 2-methylcyclohexanone-2,6,6-d_3, 77, but the carbonyl carbon was not included.[19] The isotope shifts for these cyclic ketones are shown in TABLE 4.

Norcamphor, 75, shows unusual behavior at the carbonyl carbon, C_2. An exo-d at C_3 causes an upfield shift of 0.07 ppm, while dideuteration at C_3 causes a downfield shift of -0.18 ppm. The endo-d compound was not examined. Apparently there must be a sensitive balance between the deshielding and shielding effects of deuteration adjacent to a carbonyl, such that the stereochemically distinct exo and endo deuteriums have opposite effects. There is no stereochemical dependence for isotope shifts at the saturated carbons C_4 and C_5, despite the large difference in vicinal angles between C_5 and the exo and endo positions at C_3.[53] In the camphor-3,3-d_2 molecules studied, the β-$\Delta\delta_c$ values were -0.12, -0.01, and -0.04 ppm for the 4-CH_3, 4-Cl, and 4-NO_2 compounds. It is not entirely clear whether the $\Delta\delta_c$ in the camphors and norcamphor in the study by Morris and Murray were determined from separate solutions of the labelled and unlabelled compounds, but that appears to be the case.[53] The authors do refer to spectra of binary solutions in which they were unable to resolve C_7 resonances but were able to resolve C_5 resonances in the camphors. Since the $\Delta\delta_c$ is as large as 0.06 ppm in 4-chloro-camphor-3,3-d_2, and since smaller shift differences than those given for C_7 are reported, such $\Delta\delta_c$ values must be considered as being within the probable error limits. Reference is made to

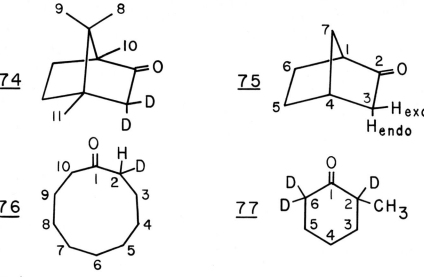

TABLE 4

Isotope Shifts in Ketones, $\Delta\delta_c$, ppm

Carbon	75-exo-d	75-d$_2$	74	76-2-d	77
C-1	0.01	-0.01	-0.01	-0.06	
C-2	0.07	-0.18	-0.12	0.364	
C-3	0.35	0.77	0.94	0.077	≤0.03
C-4	0.09	0.18	0.14	0.048	≤0.05
C-5	0.04	0.07	0.07	0.025	0.13
C-6	0.00	-0.01	-0.02	0.00	0.14 (2-CH$_3$)
C-7	0.02	0.04	0.02	-0.029	
C-8			0.01	-0.028	
C-9			0.01	0.013	
C-10			0.04	-0.29	
C-11			0.02		

scale-expanded spectra and recording at 0.04 Hz per data point,[53] but it is unlikely that these statements imply that the number of real data points were sufficient to give such digital resolution over the whole spectrum of >200 ppm range at 25.2 MHz. A spectrum of camphor and camphor-3-exo-d, has been described elsewhere, but quantitative shift differences were not given.[3,79,100]

The isotope shifts for 2-methylcyclohexanone-2,6,6-$\underline{d}_3$ and cyclodecanone-2-$\underline{d}$, are also unusual. In $\underline{77}$, the β-$\Delta\delta_c$ at C_3 is unusually small. In $\underline{76}$-2-$\underline{d}$, the long-range $\Delta\delta_c$ are unusually large and some are downfield shifts. In both cases, the unusual results may be due to $\underline{equilibrium}$ isotope effects, wherein deuteration alters the position of conformational equilibria. Anet has pointed out that the lack of an isotope effect at C_6 in $\underline{76}$-2-$\underline{d}$ is evidence for the conformational origin of the isotope shifts,[101] because low temperature nmr shows that all of the carbons $\underline{except}$ C_6 and the carbonyl are averages of two shifts from two mirror-image conformers.[102] Labelling removes the degeneracy of the equilibrium by altering nonbonded interactions and thus produces long-range shifts as well as short-range shifts. Further evidence of the perturbed equilibrium are the nearly equal shifts of opposite sign at C_5 and C_7, which is the type of splitting of resonances about an averaged position that is characteristic of perturbed degenerate equilibria (see SECTION IVA). A conformational effect may also account for the unusual behavior of $\underline{77}$,[103] by a shift in the equilibrium between chair conformers.

Other polar, unsaturated bonds that would be expected to be involved in hyperconjugation with C-H bonds on adjacent carbons are subject to either small or negative β-$\Delta\delta_c$. C_2 of the oxime of acetone is shifted upfield by only 0.08 ppm in the $\underline{d}_6$-derivative.[104] One study reported no observable β-$\Delta\delta_c$ for acetonitrile,[104] while Pohl found -0.01 ppm, based on separate measurements.[57] Pohl also reported no isotope shift for the carbonyl carbon in N,N-dimethylacetamide-$\underline{d}_9$.[57] Tulloch found no isotope shift at the carbonyl carbon in deuterium labelled methyl octadecanoates.[50]

Clearly, the β-isotope shift is reduced in these cases below the typical 0.1 ppm per ^{2}H found in alkyl and aryl systems. Since there are well-defined examples of downfield β-isotope shifts for ketones and carbocations, it seems that the reduced effects for these other cases of less electron-deficient carbons are due to the opposing action of the normal upfield shift and the hyperconjugative downfield shift. However, it must be noted that electronegative substituents also appear to reduce β-$\Delta\delta_c$ at saturated carbons.[50,94]

(B) Effects of Deuterium Bonded to Heteroatoms

Protons bound to heteroatoms are easily exchanged for deuterons, often simply by dissolution in the deuterated version of a protic solvent, such as D_2O. However, while the preparation of hetero-deuterated materials for ^{13}C nmr study is easy, the proton/deuteron exchange must be slow on the nmr time scale in order for accurate $\Delta\delta_C$ values to be determined by measuring shift differences directly from mixtures of labelled and unlabelled compounds. Slow exchange can be achieved by choice of solvent or pH, or less accurate values of $\Delta\delta_C$ may be obtained from separate solutions. Hydrogen-bonding and pK values may differ between deuterated and undeuterated samples, thereby contributing to isotope shifts. Nevertheless, the isotope shift due to deuteration at heteroatoms is useful for signal assignments, especially in substances like peptides or carbohydrates in which there are a variety of heteroatom linkages to carbon. The β-$\Delta\delta_C$ due to deuteration at nitrogen are of the same size as those produced by deuteration at carbon, but deuteration at oxygen causes a shift that is roughly double in size.

(1) Deuteration at Nitrogen. Feeney et al. showed that deuteration at amide nitrogens in peptides produced a β-$\Delta\delta_C$ of 0.10 ppm at the carbonyl.[105] Slow NH exchange was achieved by running spectra at pH 1.2 or 2.4 in 50:50 H_2O/D_2O. Under these conditions, the amide carbonyls appeared as doublets due to the isotope shift, and could be distinguished from carboxylic carbonyls, which were not split.[105] In similar fashion, Newmark and Hill showed that secondary amide carbonyls gave two peaks and primary amide carbonyls gave three when amides were dissolved in dimethyl sulfoxide, dimethylformamide, or hexamethylphosphortri-amide containing 10% H_2O and 10% D_2O.[106] Proton exchange is slow in these highly polar, aprotic solvents. In several primary amides the β-$\Delta\delta_C$ per ^2H was 0.06 ppm, with no other resolved isotope shifts. The β-$\Delta\delta_C$ at the carbonyl was 0.07 to 0.08 ppm in secondary N-aliphatic amides and 0.08 to 0.10 ppm in secondary N-aryl amides. The alkyl or aryl carbons β to the deuterium were also shifted upfield by 0.08 to 0.10 ppm. Some small γ-$\Delta\delta_C$ were observed at carbonyls β to N, but oddly, the γ-$\Delta\delta_C$ at the ortho carbons in N-aryl amides were as large as the β-$\Delta\delta_C$ for C_1, at

0.08 to 0.10 ppm. Meta and para carbons showed no resolvable shifts.[106]

Isotope shifts were determined for a few amino acids, dipeptides and related compounds by Grant et al.[107] In this study, all NH and COOH sites were completely exchanged with D_2O before ^{13}C spectra were taken in trifluoroacetic acid-d (TFA-d) and compared to spectra of the undeuterated compounds in TFA. All amine sites and carboxylate groups are protonated in this strong acid, and the amide groups are also protonated to a varying extent. Isotope shifts in the model compounds ethylamine hydrochloride and acetic acid are representative of effects in amino acids. In $CH_3CH_2ND_3^+$, the β- and γ-$\Delta\delta_c$ are 0.32 and 0.13 ppm, respectively, while in CH_3COOD, the β- and γ-$\Delta\delta_c$ are 0.27 and 0.02 ppm. The isotope shift is dependent on the number of deuterons attached to the nitrogen and the distance from the deuterated site.[107] The distance dependence is illustrated below, in the $\Delta\delta_c$ for lysine. The isotope shifts are approximately cumulative.

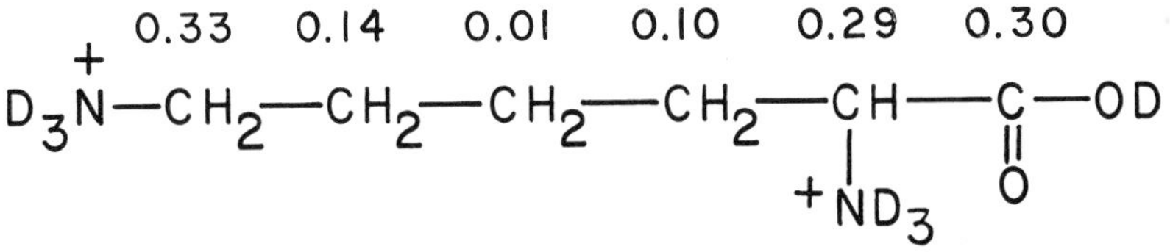

$$\overset{0.33 \quad 0.14 \quad 0.01 \quad 0.10 \quad 0.29 \quad 0.30}{\overset{+}{D_3}N-CH_2-CH_2-CH_2-CH_2-CH-C-OD}$$

with $\overset{+}{N}D_3$ and O below.

In another study, the ^{13}C chemical shifts of the amino acids glycine, alanine, and lysine were followed as a function of pH in H_2O and D_2O.[108] Intrinsic isotope shifts were not the major source of differences, rather isotope shifts were produced by isotope effects on the titration curve and, at high pH, isotope effects on pK values of the ammonium groups.

(2) Deuteration at Oxygen. It is well known that solutions of alcohols in DMSO often exhibit coupling to the hydroxyl protons in ^{1}H nmr spectra, characteristic of slow exchange. Thus, direct determination of isotope shifts is possible, but simple alcohols do not appear to have been examined for isotope shifts due to

O-deuteration, in DMSO or any other conditions. In polyfunctional compounds, especially those containing basic or acid sites, exchange may be fast even in DMSO.

The few hydroxylic systems that have been examined for isotope shifts were acidic substances or carbohydrates. The β- and γ-$\Delta\delta_c$ for acetic acid were found to be 0.27 and 0.02 ppm from separate measurements of the labelled and unlabelled acid in TFA and TFA-d.[107] Structural identification of a natural product, secalonic acid G, was aided by comparing the isotope shifts produced by deuterium exchange at the hydroxyl groups to the isotope shifts of a secalonic acid of known structure.[109] The deuterated and undeuterated substances were examined separately in CDCl$_3$. The β-$\Delta\delta_c$ at phenolic positions in tetrahydroxanthone ring systems were 0.16 to 0.17 ppm upfield, but smaller shifts, both upfield and downfield, were observed throughout the aromatic ring segment. The secalonic acids contain two tetrahydroxanthone units and each unit contained two hydroxyl functions in the acids that were studied. The hydroxyls are in a position where they can hydrogen bond to a carbonyl oxygen and tautomeric equilibria are possible, so isotopic perturbation of equilibria may be partly responsible for the shifts in the case of the secalonic acids.

secalonic acid G

Carbohydrate hydroxyl groups exchange rapidly even in DMSO. By using DMSO, low temperature (14°C), and high fields (62.8 MHz ^{13}C nmr), Gagnaire and Vincendon were able to detect a broadening in the resonances for carbons with directly bonded hydroxyl groups in 50% deuterium-exchanged mono- and disaccharides.[110] Ho, Koch, and Stuart found exchange to be fast in DMSO, so they measured isotope shifts separately in DMSO-d$_6$ or D$_2$O for O-deuterated carbohydrates and DMSO-$\underline{d}_6$ or H_2O for the unlabelled substances, using identical concentrations.[111] Upfield shifts of 0.1 to 0.2 ppm were found for ^{13}C resonances with directly bonded hydroxyls. C_3 of pyranosides, and C_2 and C_3 of pyranoses were usually shifted more than other carbons. Carbons not bonded to hydroxyls were shifted 0.00 to 0.05 ppm upfield.

The measurement and application of isotope shifts to signal assignment in carbohydrates was elegantly refined by Pfeffer, Valentine, and Parrish.[112] Unlabelled and deuterated mono- and disaccharides were used as separate solutions in H_2O and D_2O, but the resonances were measured simultaneously by using a coaxial cell arrangement, with the H_2O solution in the inner cell and D_2O solution in the outer cell. The authors called this method the differential isotope shift (DIS) technique. The isotope shifts can be measured directly from a DIS spectrum despite the rapid exchange of hydroxyl protons in aqueous solution, as shown for methyl α-D-glucopyranoside in FIGURE 3.

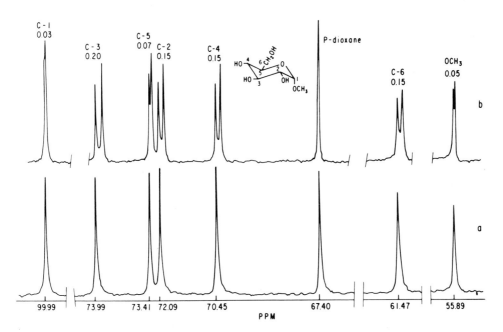

FIGURE 3. (a) 15.04-MHz proton noise decoupled spectrum of methyl α-D-glucopyranoside in H_2O; (b) DIS spectrum taken with a dual coaxial tube containing separate solutions in D_2O and H_2O. Reproduced from an article not subject to copyright.[112]

Linear regression analysis was applied to the isotope shift data from 12 D-gluco- and D-galactopyranoses and pyranosides to define isotope shift parameters.[112] The isotope shift parameters are the β- and γ-$\Delta\delta_c$ values due to O-deuterated hydroxyls that are necessary to account for the observed shifts, assuming additivity. The β-$\Delta\delta_c$ were 0.11 ppm at C_1, 0.14 ppm at C_2 through C_5, and 0.15 ppm at C_6. The smaller isotope shift at C_1 is analogous to the smaller β-$\Delta\delta_c$ at C_1 due to C-deuteration at C_2 in D-glucose (see SECTION IIIA1).[95] The parameter for γ-isotope shifts was 0.03 ppm, with no statistically significant variation due to axial vs. equatorial hydroxyls, except that a parameter of 0.06 ppm was needed for the effect of a trans hydroxyl at C_1. These isotope shift parameters were then applied to the prediction of isotope shifts in other mono- and disaccharides and derivatives. Comparisons with observed isotope shifts allowed the assignment of otherwise difficultly distinguishable resonances and resulted in several reassignments of literature data.[112]

(C) Isotope Effects of Nuclei Other than Deuterium

Two- and three-bond isotope shifts at carbon due to isotopic substitutions other than $^1H/^2H$ are expected to be smaller than ^2H-induced shifts, like the smaller one-bond effects discussed in SECTION IIB. The available data confirm this, but are too few to make any other generalizations. In 1,1'-dimethylferrocene, an average β-$\Delta\delta_c$(^{13}C) of 0.003 ppm was found for the $^{12}C/^{13}C$ effect, which is one-sixth of the α-shift but also within the limits of experimental error.[87] In cyclobutanone, ^{13}C at C_3 produces a downfield isotope shift of -0.012 ppm at C_1, the carbonyl carbon.[89] In n-propyl benzoate, a β-$\Delta\delta_c$(^{18}O) of 0.005 ppm was found at the central methylene carbon of the propyl group for $^{16}O/^{18}O$ substitution at the alkoxy oxygen of the ester.[86] This β-$\Delta\delta_c$(^{18}O) is about one-sixth of the α-$\Delta\delta_c$(^{18}O) of 0.032 ppm. No three-bond effects on ^{13}C shifts have been reported for isotopic substitutions other than $^1H/^2H$.

IV. LONG-RANGE AND DOWNFIELD ISOTOPE SHIFTS

(A) Equilibrium vs. Intrinsic Isotope Effects

Only two examples of downfield shifts produced by heavy atom substitution were cited in the review of isotope effects on nmr resonances by Batiz-Hernandez and Bernheim, published in 1967.[18] Kanazawa, Baldeschwieler and Craig had found a downfield shift for the ^{1}H signal of cis-1,2-difluoroethane-1-d,[113] and successive deuterium substitution had been found to produce downfield shifts of the remaining protons in the ammonium ion.[114] The latter result was attributed to an isotope effect on solvation rather than an intrinsic isotope shift.[18,114] Since that review, the only report of a one-bond, α-isotope shift to lower field was in the case of perdeuterated metallocenes of paramagnetic character.[76] However, there are now several reliable cases of downfield shifts for two-bond and longer-range effects, although it is clear from the examples cited in SECTION III that upfield shifts are far more common.

The intrinsic isotope effect on chemical shifts is ordinarily also a short-range phenomenon, in contrast to equilibrium isotope effects which may affect positions considerably removed from the site of substitution. No examples of isotope shifts occurring over more than three bonds were cited by Batiz-Hernandez and Bernheim.[18] The intrinsic effect is usually largest for a nucleus directly bonded to the isotopic atom and diminishes to virtually zero when the nuclei are separated by more than three bonds.[115] Anet observed that no intrinsic isotope effect on ^{13}C chemical shifts could be observed as long as there were at least four bonds separating the nuclei, even in a case where steric compression produces a long-range isotope effect for a ^{1}H resonance.[101,103] However, Anet also suggested that long-range intrinsic isotope effects would be expected in conjugated (resonance stabilized) anions or cations, or in molecules where the deuteron is close in space to the observed nucleus.[101] A number of examples are now known and cited below, of intrinsic, long-range isotope shifts including shifts of ^{13}C resonances. Several of these are in cations or anions, but others are not associated with ions or with spatial proximity between the nuclei.

It is important to distinguish between intrinsic isotope effects and possible equilibrium isotope effects, especially when considering long-range or downfield chemical shifts. Equilibrium isotope effects are readily detected in some cases of isotopic perturbation of degenerate equilibria, because the isotopic perturbation will lift the chemical shift degeneracy and two resonances will be observed for formerly equivalent nuclei. This type of equilbrium isotope effect on chemical shifts is particularly striking in the cation rearrangements studied by Saunders, with isotopically induced splittings as large as 105 ppm for ^{13}C resonances.[78] Much smaller isotopic splittings are induced by isotopic perturbation of degenerate conformational equilbria in neutral molecules. For instance, the ^{13}C resonances of the 3,3-methyl groups in 1,1,3,3-tetramethylcyclohexane, labelled with one 1-CD_3, differ by only 0.184 ppm.[103] Diagnostic of such cases is the split of the resonances about the resonance position for the unlabelled compound; one peak will be shifted upfield, the other downfield by an equal amount in the absence of any additional intrinsic effect. Isotopic perturbation of a degenerate conformational equilbrium is probably responsible for the long-range and downfield isotope shifts seen in cyclodecanone,[62] as discussed in Section IIIA2c.

Chemical shifts induced by isotope effects on equilibria are more difficult to distinguish from intrinsic isotope shifts when the perturbed equilibria are not degenerate. For instance, Anet has postulated that the unusual isotope shifts in 2-methylcyclohexanone[19] are due to an isotope effect on a conformational equilibrium,[103] where the two conformers are nondegenerate chair conformers (see SECTION IIIA2c). For nondegenerate equilibria, the isotopically induced splitting of resonances about an averaged position will not be seen. To identify such equilibrium effects with certainty, it may be necessary to carefully study the temperature dependence of the isotope shifts and compare it with the temperature dependence in systems where conformational effects are not possible. In principle it is possible in most flexible molecules that isotopic substitution would perturb conformational equilibria, but this perturbation is unlikely to be significant unless there are important contributions from conformations with significant nonbonded (steric) or hyperconjugative interactions at the site of substitution.

Isotopic subtitution may also perturb intermolecular inter-actions, such as hydrogen-bonding, or other forces of solvation, complexation or association. For instance, downfield isotope shifts were reported for proton resonances in partially deuterated benzenes.[116] However, Savitsky studied $\underline{d}_4$-nitrobenzenes and benzene-$\underline{d}_5$ and found that while downfield isotope shifts occurred in neat liquids, no significant isotope effect could be detected at infinite dilution in CCl_4.[117] Apparently the isotope shifts at higher concentrations are due to isotope effects on the degree of association between benzene molecules.[117] In a case of complex-ation with an europium shift reagent, a doubling of all ^{1}H resonances was seen for alcohols which were run as mixtures of unlabelled and $\underline{d}_1$-labelled (at the hydroxyl-bearing carbon) compounds.[118] The isotope effects in the presence of the shift reagent were attributed to a greater association constant between the deuterated compound and the shift reagent.[118] In a more unusual case, the two stereoisotopomers 78 and 79 gave different methoxy proton resonances in a complex with $Eu(fod)_3$.[119] Isotope effects on hydrogen bonding or tautomeric equilibria may be

78   79

involved in long-range and downfield shifts observed in secalonic acids when phenolic hydrogens are replaced by deuterium (see SECTION IIIB2).[109] Obviously, it may be very difficult to rule out at least some contribution to isotope shifts from isotope effects on intermolecular interactions, unless the spectra are obtained for the gas phase.

(B) Hyperconjugation and Isotope Shifts

(1) Long-Range Isotope Shifts. The first long-range, down-field isotope shift was reported by Traficante and Maciel in 1965, for the ^{19}F resonance in p-fluorotoluene-α,α,α-$\underline{d}_3$, 82.[120] The ^{19}F isotope shift for 82 was -0.01 ppm, downfield of p-fluorotoluene. No isotope shift was observed for m-fluorotoluene-α,α,α-$\underline{d}_3$. No isotope shift was detected for either para or meta ^{13}C resonances in toluene-α,α,α-$\underline{d}_3$.[55] Small upfield shifts were detected for ^{19}F in the monodeuterio fluorobenzenes: 0.011 ppm for p-F and 0.008 ppm for m-F, which are long-range effects, and 0.285 ppm for o-F, which is a γ-isotope shift.[121] No long-range effect was observed in benzene-1-$\underline{d}$ at the para carbon, and the γ- and β-$\Delta\delta_c$ for the meta and ortho carbons were 0.011 and 0.110 ppm, respectively.[71] Clearly the ^{19}F chemical shift is more sensitive than the ^{13}C chemical shift to isotopic substitution in the molecule, even though the fluorine is one bond more distant from the site of substitution. This greater sensitivity is characteristic of fluorine nmr shifts in general. For instance, p-^{19}F chemical shifts are more than seven times more sensitive in Hz ($\sim$2x in ppm) than are p-^{13}C shifts in phenylcarbenium ions.[122,123]

Long-range, downfield shifts were observed in ^{19}F resonances of a series of p-fluorophenylcarbenium ions, by Timberlake, Thompson, and Taft.[43] The downfield shifts associated with deuteration of a methyl group at the carbenium center were an order of magnitude larger than for p-fluorotoluene-$\underline{d}_3$. For instance, the $\Delta\emptyset_F$ for several p-fluorophenyl systems, including those of Taft et al., are listed in TABLE 5, in the order of the ^{19}F chemical shifts, $\emptyset_F$, of the undeuterated compounds. Clearly, this order is nearly the same as the order of the isotope shift ($\Delta\emptyset_F$) per CD$_3$ group. In fact, for the ions 86-91', Taft demonstrated a linear correlation between the isotope shift per CD$_3$ group and the fluorine chemical shift.[43]

TABLE 5

^{19}F Isotope Shifts for p-Fluorophenyl Derivatives

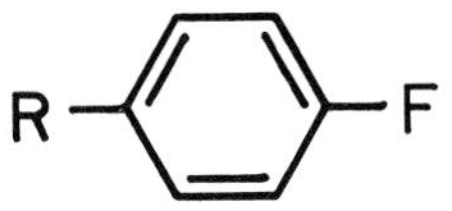

R — ⟨benzene ring⟩ — F

Compound	R	$\emptyset_F$ [a]	$\Delta\emptyset_F$ [b]	Ref.
80	NHCD$_3$	130.1	0.033	44
81	OCD$_3$	124.6	0.045	44
82	CD$_3$	118.7	-0.01	120
83	D	113.2	0.011	121
84	N(CD$_3$)$_3^+$	111.1	0.0	43
85	COCD$_3$	106.7	0.0	43
86	C(CD$_3$)NH$_2^+$	95.3	-0.03	43
87	C(CD$_3$)OH$^+$	80.7	-0.11	43
88	C≡C-C(CD$_3$)$_2^+$	77.5	-0.28	43
89a	C(CD$_3$)CH$_3^+$	60.9	-0.230	44
89b	C(CD$_3$)$_2^+$	60.9	-0.460	43,44
90a	CHCD$_3^+$	47.0	-0.324	43,44
90b	CHCD$_2$H$^+$	47.0	-0.204	44
90c	CHCH$_2$D$^+$	47.0	-0.092	44
90d	CDCH$_3^+$	47.0	0.00	44
91'	C(CD$_3$)CF$_3^+$	25.6	-0.40	43

[a] Fluorine nmr shift of the undeuterated substance in ppm from CFCl$_3$. Positive values are upfield.
[b] Isotope shifts in ppm. Positive values are upfield shifts.

Similar long-range isotope shifts have been observed for ^{13}C resonances of charge-delocalized organic ions. Servis and Shue reported downfield shifts of -0.2 ppm at the ortho and para carbons in the $\underline{d}_6$-phenyldimethylcarbenium ion, 91.[40] These $\Delta\delta_C$ were later determined with greater precision to be -0.16 and -0.22 ppm for ortho and para, respectively.[44] An upfield shift of 0.20 ppm was found for the para carbon in the $\underline{d}_5$-α-methyl-α-neopentyl anion, 92.[44]

$CD_3 \overset{+}{\underset{C}{\diagup}} CD_3$

$CD_3 \overset{-}{\underset{C}{\diagup}} CD_2 \diagup$

91

92

Since the long-range isotope effects and downfield shifts reported for the p-fluorophenyl- and phenyl-substituted ions are unusual, it is important to establish that these are intrinsic isotope effects, not equilibrium effects. It could be suggested that an isotope effect perturbing an equilibrium between free ions and ion pairs, or between different ion-paired species, could produce the observed isotope shifts. In the case of the carbocation 89b, this possibility was ruled out by the lack of any significant dependence of the isotope shift on temperature, concentration, counterion, or acid medium.[44] For the carbanion 92, the isotope effect was independent of temperature and of added tetramethylethylenediamine, which would complex with the lithium counterion.[44]

Several lines of evidence indicate that the long-range isotope effects on chemical shifts arise from the hyperconjugative interactions of the methyl groups and are transmitted via the π system (see also Section IB). Taft et al. demonstrated that the magnitude of the isotope effect on the ^{19}F shift in methyl-p-fluorophenylcarbenium ions was related to the electron demand on the methyl group, as indicated by the linear dependence on the chemical shift, $\emptyset_F$.[43] For instance, the isotope effect in 89a, where two methyl groups stabilize the charge, is smaller than in 90a, where only one methyl is present. The isotope shift is very small for toluene, 82, where the hyperconjugative interaction is minimal. An "inductive" origin of the isotope effect is ruled out by the lack of an isotope effect on the ^{19}F shift in 90d, where a nonhyperconjugating α-hydrogen is replaced by deuterium. A "steric" origin of the isotope effect is ruled out by the isotope effect in 88, where a triple bond has been interposed between the side chain and the phenyl ring. That an isotope effect on ^{13}C chemical shifts in 91 is observed at the ortho and para positions but not at the meta position, strongly suggests transmission via the π system. Finally, the long-range isotope effects for 91 and 92 are in the opposite direction: downfield isotope shifts at ortho and para carbons in the cation 91, and an upfield shift at the para carbon in the anion 92.

The effects of substituting entire CD_3 groups for CH_3 groups are additive, as may be seen for 89a and 89b in TABLE 5. This is the expected result for two methyl groups equally involved in hyperconjugation. However, each successive deuterium substitution produces a larger, long-range, downfield shift in the ^{19}F signal of the methyl-p-fluorophenylcarbenium ion, 90a,b,c. The nonadditive behavior was attributed to unequal populations of the possible methyl conformations for partially deuterated methyl groups, so that each C-H(D) bond is not equally involved in hyperconjugation.[44] The preferred conformers are those in which C-H bonds are better aligned for hyperconjugation than C-D bonds.

(2) Downfield β-Isotope Shifts. Servis and Shue have used the isotope shift due to hyperconjugating β C-D bonds as an indicator for the mode of charge delocalization in carbocations.[40] β-Deuterium isotope effects on ^{13}C resonances of cation centers are downfield isotope shifts if the ions are localized, static ions. The downfield shifts are due to better electron release by C-H than C-D. For example, the $\beta-\Delta\delta_c$ is -1.4 ppm for the t-butyl cations, 93, and -0.4 ppm for the methylcyclopentyl cation, 94. The smaller, downfield $\beta-\Delta\delta_c$ in acetone-$\underline{d}_6$ and other ketones is probably also due to the hyperconjugative effect (see SECTION IIIA2c).

For classical charge-delocalized ions, such as 91, and the methylcyclohexenyl cation, 95, or cyclopropyl-stabilized ions, the $\beta-\Delta\delta_c$ is zero. In these cases, the downfield, hyperconjugative isotope shift is balanced by the normal upfield $\beta-\Delta\delta_c$ (termed an inductive isotope effect by Servis and Shue). The demand for hyperconjugative stabilization is less in the charge-delocalized ions, and electron density released from the β C-H(D) bonds will be spread over the π system.

For nonclassical π- or σ-bridged ions, upfield β-isotope shifts are observed and these are larger than normal $\beta-\Delta\delta_c$

apparently due to isotopic perturbation of resonance.[14,40]   For example, the π-bridged 7-methyl-7-norbornenyl cation, <u>96</u>, exhibits a β-$\Delta\delta_C$ of 0.8 ppm, and the σ-bridged 2-methyl-2-norbornyl cation, <u>97</u>, has a β-$\Delta\delta_C$ of 2.2 ppm.

93   94   95

96   97

(C)   Other Long-Range or Downfield Shifts

Most long-range and/or downfield isotope shifts can be ascribed to either equilibrium effects or the effects of isotopic substitution at bonds involved in hyperconjugation, as discussed in the preceding two sections.   However, some examples remain which do not fit those explanations.

Some reports of long-range or downfield shifts are doubtful because the shifts are probably within the limits of accuracy of the measurements.   In the following four instances, the data were obtained from separate solutions, not binary mixtures, and should be reconfirmed: (i)   a downfield shift of −0.05 ppm at C_5 in benzonitrile-2-$\underline{d}$[75];   (ii)   a γ-$\Delta\delta_C$ of −0.04 ppm in methyl hexadecanoate-2,2-$\underline{d}_2$[60] (note that an upfield γ-$\Delta\delta_C$ was found in methyl octadecanoate-2,2-$\underline{d}_2$[50]);   (iii) a four-bond $\Delta\delta_C$ of 0.01 ppm in lysine due to N-deuteration[107]; and (iv) a γ-$\Delta\delta_C$ of −0.01 ppm at C_1 in two camphors and norcamphor due to dideuteration at C_3.[53]

An upfield isotope shift of 0.011 ppm was noted in TABLE 5 for fluorobenzene-4-d and this, plus the similar effect for fluorobenzene-3-d,[121] was attributed to the high sensitivity of the ^{19}F chemical shift. However, recently an isotope shift of -0.004 ppm was reported for the tritium nmr of $[1,4-^3H_2]$benzene compared to $[^3H]$benzene.[124] This unusual result is possibly related to results for ^{1}H nmr of partially deuterated benzenes, which displayed downfield isotope shifts at high concentration but no significant shifts at low concentration, indicative of isotope effects on intermolecular interactions (see SECTION IVA).[117] Unusually large four-bond effects on ^{19}F resonances are found in fluoroacetamides due to N-deuteration.[125] The isotope shift is 0.265 ppm for fluoroacetamide-N,N-d_2 in DMF, diminishing to 0.119 ppm for the difluoro compound and to 0.065 ppm for trifluoroacetamide-N,N-d_2.[125]

Long-range, downfield shifts for ^{19}F were observed by Mitchell and Phillips in the cis- and trans-stilbenes, 98 and 99, for substitution of a vinylic hydrogen by deuterium.[126] A nearly constant effect of -0.019 ppm was noted in the cis-series, 98, but the isotope shift varied from -0.005 ppm for NMe$_2$-99 to -0.013 for NO$_2$-99. The isotope shifts were attributed to the ability of the deuterated compounds to become more nearly co-planar because of the effectively smaller size of the C-D bond, i.e., a C-D bond has a smaller steric requirement than a C-H bond. Interestingly, in both 98 and 99, when X=F the second fluorine is shifted upfield in the deuterated compound compared to the unlabelled difluoro-stilbene.[126]

98   99

More recently, Anet and Dekmezian described upfield isotope shifts for protons which were close in space to deuterons, but separated by several bonds.[101] This represents a direct steric isotope effect on the observed nucleus. However, the alkyl carbons to which the protons were attached were not shifted, and normal α, β, and γ shifts were reported.[101]

Downfield shifts were also found in some alkane rings, as described in SECTION IIIA1. These examples are a $\gamma - \Delta\delta_c$ of −0.012 ppm for cyclopentane-1-$\underline{d}$, a four-bond $\Delta\delta_c$ of −0.014 in cyclo-heptane-1-$\underline{d}$, and a $\Delta\delta_c$ of −0.037 at C_4 in norbornane-1-$\underline{d}$.[68] An order of magnitude larger downfield $\gamma - \Delta\delta_c$ were found for the cyclopentenyl (68) and cyclohexenyl (69) cations (and possibly the tropylium ion 70, see SECTION IIA2), and were ascribed to isotopic perturbation of resonance.[14] In these cations, the C−H(D) bonds are orthogonal to the π system, so they cannot be involved in hyperconjugation.

V. ISOTOPE EFFECTS ON SPIN-SPIN COUPLING

In addition to the effects of isotopic substitution on chemical shifts, isotope effects on spin-spin coupling constants are also possible. A primary isotope effect on a coupling constant is an effect on the reduced coupling constant, $J_{AX} / \gamma_A \gamma_X$, upon isotopic substitution of nucleus A or X. The expected relationship between the coupling of a nucleus, X, with hydrogen or deuterium is given by the ratio of the magnetogyric ratios for hydrogen and deuterium:[127]

$$J_{XH} = (\gamma_H / \gamma_D) J_{XD} = 6.5144 \ J_{XD} \tag{2}$$

Experimental deviations from this relationship indicate the possibility of a primary isotope effect, but relaxation effects due to the deuterium quadrupole moment could also result in smaller than expected J_{XD} values.[61,128] A secondary isotope effect on a coupling constant is a change in J_{AX} when a nucleus other than A or X is isotopically substituted. Isotope effects may also be classified as negative or positive, depending on the direction of the effect. A negative isotope effect is one which diminishes the absolute magnitude of the reduced coupling constant, or in terms of coupling to hydrogen or deuterium, results in a smaller J_{XD} than expected.

Both primary and secondary isotope effects on coupling constants are known. Most of the strong evidence for such effects is for inorganic molecules, probably due in part to the occurrence of large coupling constants in some such species, and possibly due to greater isotope effects on intermolecular interactions than in hydrocarbons. With coupling constants as with isotope shifts (SECTION IVA), it is important but perhaps very difficult to distinguish between intrinsic isotope effects and those induced by isotope effects on solvation or other intermolecular interactions. One difficulty in demonstrating primary isotope effects is that the experimental uncertainty in measuring J_{XD} becomes magnified in making a comparison with J_{XH}. For example, if the uncertainty is $\pm$ 0.1 Hz for J_{XD}, the uncertainty in a J_{XH} calculated from J_{XD} is ±0.65. If the error in J_{XH} (obs) is ±0.1 Hz, the primary isotope effect must be larger than 0.75 Hz in J_{XH} (calc) to be outside the range of experimental error. The difference J_{XH} (calc) $-$ J_{XH} (obs) will be referred to herein as (ΔJ), the isotope effect on the coupling constant, eq. 3.

$$\Delta J = J_{XH}(\text{calc}) - J_{XH}(\text{obs}) \tag{3}$$

Measurements of secondary isotope effects do not suffer from this difficulty because the same nuclei are involved in both couplings. For secondary effects, ΔJ will be defined by eq. 4.

$$\Delta J = J_{AX}(\text{heavier isotopomer}) - J_{AX}(\text{lighter isotopomer}) \tag{4}$$

(A) Inorganic Compounds

No secondary isotope effect was found in the ammonium ion for $^{14}N-^{1}H$ coupling in the series $NH_4^+/NH_3D^+/NH_2D_2^+/NHD_3^+$,[129,130] although the proton resonances were isotopically shifted.[130] Similarly, the coupling constants $J(^{15}N-^{1}H)$, $J(^{15}N-D)$, and J_{HD} were invariant at 61.8 ± 0.5, 9.45 ± 0.20, and 1.54 ± 0.12 Hz, respectively, in the ammonia series $NH_3/NH_2D/NHD_2/ND_3$.[131,132] Thus, despite isotope shifts for both ^{15}N and ^{1}H in ammonia, no primary or secondary effects on coupling were observed.

On the other hand, both positive and negative primary isotope effects were found for one-bond coupling to ^{31}P that were well outside experimental error.[65] For instance, a positive effect of $J_{PH}(calc) - J_{PH}(obs)$, or $\Delta J = 2.0 \pm 0.8$ Hz was found for $C_6H_5PH_2/C_6H_5PHD$, as well as positive effects for other three-coordinate phosphorus compounds. A negative effect of $\Delta J = -3.9 \pm 0.8$ Hz was found for deuterium substitution in four-coordinate phosphorus compounds, such as dimethylphosphonate, $(CH_3O)_2P(O)H/(CH_3O)_2P(O)D$. However, another study of the isotope effect on J_{PH} demonstrated that the ΔJ in dimethylphosphonate was dependent on solvent, concentration, and whether the compounds were present in the pure state or mixed with the deuterated compounds.[133] This study ascribed the isotope effects to a difference in hydrogen-bonding capability of the deuterated and nondeuterated species. A study of liquid phosphine and its isotopomers disclosed a positive isotope effect for the primary isotope effect, i.e., the reduced coupling constant for P-D was larger than the P-H reduced coupling.[134] However, the magnitude of both couplings diminished with successive deuteration in the series $PH_3/PH_2D/PHD_2/PD_3$, indicating a negative sign for the secondary isotope effect. A negative secondary isotope effect was also found for $^{19}F-^{31}P$ coupling in H_2PO_2F/D_2PO_2F, where J_{PF} dropped 3.5 ± 0.2 Hz, from 1033.5 to 1030.0 Hz.[38] The primary isotope effect, ΔJ, was negative for H_2PO_2F/D_2PO_2F.[38] From these four studies of phosphorus compounds, the primary isotope effects in four-coordinate compounds are negative and the primary isotope effects in three-coordinate compounds are positive. The secondary isotope effects on J_{PH} are negative in phosphine[134] and H_2POOH,[65,135] but positive in $(C_6H_5)PH_2$.[65]

Studies of other inorganic compounds also indicate a mixed set of results, with both positive and negative signs for secondary isotope effects. Negative secondary isotope effects were found for tetrahydroborate anions, with no significant primary effect.[136] Negative primary and secondary isotope effects were observed in $HSiF_3/DSiF_3$.[38] The secondary effect was negative for SiH_3I/SiH_2DI,[137] but positive for $SeH_2/SeHD$.[138] The isotope effect on J_{HD} in $SiH_3D/SiHD_3$ was negative.[139] No secondary isotope effect on J_{SnH} was observable for $(C_6H_5)_2SnH_2/(C_6H_5)_2-SnHD$.[137] In hydrogen cyanide, the small secondary effect on $^{13}C-^{15}N$ coupling is positive, as J_{CN} increases from 18.5 ± 0.10 Hz

in HCN to 18.8+0.10 Hz in DCN.[69]   The primary isotope effect for
coupling to ^{13}C in HCN/DCN is negative, but within the error
limits.

In all the inorganic compounds cited here, positive isotope
effects on coupling constants are observed only in non-tetra-
coordinate structures.   Most primary and secondary effects are
negative, i.e., coupling constants in deuterated compounds are
smaller than expected.   A negative primary isotope effect was
predicted theoretically for the hydrogen molecule[6] and for
methane,[140] based on the bond shortening accompanying isotopic
substitution.   The positive isotope effects may be associated with
more important bond angle changes in di- and tricoordinate molecules.[38]   The largest isotope effects amount to only a few Hz,
even in systems with very large coupling constants.   Differential
solvation for labelled and unlabelled molecules may be important
but have been little explored.   Besides the study of phosphonates[133] previously mentioned, the only other instance where
solvent effects were examined experimentally was the report that
$LiBH_4$ in tetrahydrofuran and $NaBH_4$ in water were subject to very
similar isotope effects.[136]

(B)   Organic Compounds

The first report of isotope effects on coupling constants in
organic molecules indicated that the $^{13}C-^1H$ coupling constant is
about 4 Hz larger in a $-CH_2D$ or $-CHD-$ group than in $-CH_3$ or $-CH_2-$,
for instance, in toluene.[37]   This report was corrected shortly
afterwards.[130]   Muller and Birkhahn found J_{CH} = 125.75+0.2 Hz in
toluene-α-d,[129] and Fraenkel and Burlant measured it as
125.9 Hz,[130] only very slightly smaller than J_{CH} = 126.0 for
toluene.[37,141]   A later study put the value at 125.4+0.2 Hz.[38]
This small, negative isotope effect that is barely significant in
relation to experimental error is typical of primary and secondary
isotope effects on coupling constants in organic molecules.   The
tendency is toward negative isotope effects, smaller than expected
couplings following isotopic substitution, but most individual
examples are within the range of experimental error.

Pohl et al. determined J_{CD} values for many perdeuterated
organic solvents and compared them to J_{CH} values for the
unlabelled compounds.[57]   The precision of the J measurements was

not high ($\pm$0.6 Hz), so only relatively large ΔJ effects could have been reliably detected in individual cases. Pohl found that the observed J_{CD} values fit the calculated line $J_{CD} = (\gamma_D/\gamma_H)J_{CH}$ with an average deviation of about $\pm$1 Hz, indicating that there is no systematic isotope effect which increases or decreases with the magnitude of the coupling constant. In a similar though less extensive study of deuterated solvents, Gold et al. measured coupling constants with somewhat greater precision.[2] In seven out of eleven cases the $\Delta J = (6.5144\ J_{CD} - J_{CH})$ were negative, two were zero and two were positive, but in all cases except chloroform the ΔJ values were within the range of experimental error. The J_{CH} for $CHCl_3$ is 208.8$\pm$0.2 Hz, J_{CD} for $CDCl_3$ is 31.9$\pm$0.1 Hz, and ΔJ is -1.0$\pm$0.9. Thus, on average the primary isotope effect appears to be negative but very small.[2]

Gas phase ^{13}C nmr measurements of methane and its deuterium isotopomers indicated no primary isotope effect on the $^{13}C^{-1}H$ coupling, within experimental error.[17] However, a small, negative, secondary isotope effect is evident in J_{CH} and possibly in J_{CD}. These results are shown in TABLE 6. Actually, the reduction in J_{CD} values with successive deuterium substitution may also be considered a primary isotope effect as well, because the deuterium is of course directly involved in the coupling that is measured. The net result in comparing CD_4 to CH_4 is a primary isotope effect, ΔJ, that is probably significant at -0.9$\pm$0.7 Hz.

TABLE 6
Coupling Constants for Methane Isotopomers[17]

	J_{CH}, Hz	J_{CD}, Hz	6.5144 J_{CD}
CH_4	125.3$\pm$0.1		
CH_3D	125.3$\pm$0.4	19.3$\pm$0.3	125.7$\pm$2.0
CH_2D_2	124.9$\pm$0.3	19.2$\pm$0.2	125.1$\pm$1.3
CHD_3	124.5$\pm$0.2	19.1$\pm$0.1	124.4$\pm$0.6
CD_4		19.1$\pm$0.1	124.4$\pm$0.6

Günther's high-field ^{13}C nmr study of monodeutero cyclo-alkanes that was so revealing of isotope shifts (see SECTIONS IIA1 and IIIA1) did not find a significant primary isotope effect on coupling constants in cyclopropane or cyclohexane.[68] Field strength does not affect signal separation due to coupling, so the error in J_{CD} values was still about ± 0.1 Hz. The chief limitation to more precise values are the typically broader lines for carbons bound to deuterium. Another study examined $^{13}C-^{119}Sn$, $^{13}C-^{117}Sn$, $^{13}C-^{1}H$, and $^{13}C-^{2}H$ couplings in the tetramethyltin series $Sn(CH_3)_{4-n}(CD_3)_n$ and $^{1}H-^{2}H$ in some other partially deuterated tetramethyltin isotopomers.[39] No significant primary or secondary isotope effects on coupling constants were found.

Fraser, Pettit and Miskow reported an apparently significant negative isotope effect on the geminal $^{1}H-^{1}H$ coupling in the methylene groups of benzyl methyl sulfoxide, $C_6H_5CH_2S(O)CH_3$.[142] The ΔJ of -0.46 ± 0.2 Hz was obtained from a ^{1}H nmr spectrum of a 1:3 mixture of benzyl methyl sulfoxide and the stereoselectively labelled α-deuterio derivative. The possibility that nuclear quadrupole relaxation was affecting the J_{HD} was ruled out by finding the same splitting and linewidths at two temperatures. However, an apparently larger ΔJ of -1.6 Hz for the geminal coupling in neopentyl alcohol, $(CH_3)_3CCH_2OH$, was ascribed at least in part to relaxation effects on J_{HD} as indicated by a broader linewidth. The J_{HH} in neopentyl alcohol was measurable through the use of a chiral shift reagent.[142]

In regard to geminal $^{1}H-^{1}H$ coupling and the possible influence of quadrupole relaxation, it is interesting to consider two papers on tritium-labelled compounds by Elvidge et al.[61,143] Tritium does not have a quadrupole moment, so a relaxation effect can not be a factor. They calculate J_{HH} values from observed J_{HT} and observed J_{HD} couplings, using the magnetogyric ratios. Most of the J_{HD} values were obtained from the literature, but both J_{HD} and J_{HT} values were measured for acetone and dimethyl sulfoxide in this study. For acetone, J_{HH} is calculated to be 14.32 ± 0.03 from J_{HT} and 14.33 ± 0.07 Hz from J_{HD}. The calculated J_{HH} values for DMSO are 12.41 ± 0.04 and 12.12 ± 0.26 Hz from J_{HT} and J_{HD}, respectively. In neither case is an isotope effect significant, although it is nearly so for DMSO, with the ΔJ_{HH} positive, at 0.29 ± 0.30, for substitution by the heavier nucleus. Several additional ΔJ_{HH} values calculated from the data of Elvidge et al.

are listed in TABLE 7.[61] Certainly there is no clear pattern of significant isotope effects on the geminal coupling constants. The tritium isotope effect,[143] ΔJ_{HH}, on geminal coupling in benzyl methyl sulfoxide is only -0.11, compared to -0.46 for the deuterium effect[142] described in the preceding paragraph.

TABLE 7

Apparent Isotope Effects on J_{HH} and J_{CH}

Compound[a]	$\Delta J_{HH}(^2H,{}^3H)$	$\Delta J_{CH}(^2H)$	$\Delta J_{CH}(^3H)$
$C\underline{H}_3COCH_3$	-0.01+.10		
$C\underline{H}_3S(O)CH_3$	0.29+.30		
$C\underline{H}_3CN$	-0.82+.36	-0.83+0.87	-0.24+.10
$PhCOC\underline{H}_3$	0.02+.31	-1.94+2.10	0.24+.09
$CHCl_3$		-0.79+0.96	-1.45+.07
$C\underline{H}_2(CO_2Et)_2$		-1.01+2.12	-0.18+.12
$NaO_2CC\underline{H}_3$	-0.18+.33	-0.02+0.83	0.18+.11
$C\underline{H}_3NO_2$	-0.36+.30		

[a]Site of isotopic substitution is underlined.

Also listed in TABLE 7 are several values for primary isotope effects on the one-bond coupling to carbon. Again these are calculated from the data of Elvidge et al. for both deuterated and tritiated compounds.[61] While all of the deuterium isotope effects, $\Delta J_{CH}(^2H)$, are negative, they are also all within the range of experimental error. More importantly, the tritium isotope effects are mostly smaller in magnitude, with some negative and some positive. If the apparent trend toward negative deuterium isotope effects on couplings were due to real isotope effects arising from the greater mass of the deuterium nucleus, the tritium isotope effects should all be larger and in the same direction. That they are not suggests that any reduction in J_{CD} values may be due to deuteron relaxation.[61,143] The exception may be chloroform, and even in this case it is possible that the effect is not intrinsic but due to an isotope effect on solvent interactions, since the C-H bond is relatively polar.

VI. CONCLUSIONS

In the preceding sections the effect of isotopic substitution on ^{13}C chemical shifts and on spin-spin coupling constant have been discussed in terms of applications and variation with structure. The intrinsic isotope shift, especially the two- and three-bond effects of deuterium substitution, is likely to grow in importance as a tool for signal assignment. The isotope shift also provides a means to locate sites of isotopic incorporation and to quantify the extent of substitution, which will be especially valuable for $^{16}O/^{18}O$ substitution. The observation of long-range and downfield isotope shifts may lead to a greater understanding of delocalized bonding involving C-H(D) bonds.

Attempts to generalize about the relation of isotope effects to structural variables are hindered by three factors: (i) there is a dearth of accurate measurements obtained from binary solutions containing both labelled and unlabelled material; (ii) the effects are small so that errors are relatively large; and (iii) insufficient attention has been paid to distinguishing between intrinsic isotope effects and possible equilibrium isotope effects. The first difficulty can be removed by simply using binary solutions or the dual cell technique of Pfeffer.[112] Increased use of spectrometers with high fields and high digital resolution (data points per Hz) will improve the precision of measurement of isotope shifts, although measurement of coupling constants will not be significantly affected. The problem of possible equilibrium isotope effects is also a promise: the recognition that the isotopic perturbation of conformational equilibria and solute-solvent interactions may influence chemical shifts and coupling constants opens up these measurements as a probe of subtle intra- and intermolecular interactions.

In regard to generalizations about intrinsic isotope shifts, the most significant new developments since the review of Batiz-Hernandez and Bernheim[18] are the observations of long-range and downfield isotope shifts (SECTION IV). Theoretical treatment of these observations will require consideration of the effect of isotopic substitution on delocalized electronic wavefunctions. Jameson[23] addressed the theory of isotope shifts in relation to five empirical generalizations noted by Batiz-Hernandez and Bernheim.[18] These five generalizations, as restated by Jameson, are commented on below with regard to the observations in ^{13}C nmr discussed in the present paper.

(1) "Heavy isotopic substitution shifts the nmr signal of a nearby nucleus toward a higher magnetic field."[18,23] Shielding is still the rule for one-bond isotope shifts, except in the case of some paramagnetic metallocenes,[76] and is also the general rule for two-bond effects in saturated frameworks. However, deshielding may occur for two-bond effects in carbocations and at carbonyl carbons, or for longer range effects.

(2) "The magnitude of the isotope shift is dependent on how remote the isotopic substitution is from the nucleus under observation."[18,23] Attenuation with distance or the number of intervening bonds applies to saturated and most unsaturated systems. Isotope shifts have been observed for protons of alkyl groups which were close in space to deuterons and separated by several bonds, but the alkyl carbons were not shifted.[101] The generalization does not hold for long-range effects transmitted via resonance in conjugated systems.

(3) "The magnitude of the shift is a function of the resonant nucleus reflecting the differences in the range of chemical shifts observed for the nuclei."[18,23] This generalization was not examined in detail here because this review focused on a single nucleus. However, some cases have been cited wherein it is clear that nuclei such as ^{19}F and ^{119}Sn which have larger ranges of chemical shifts also experience larger isotope shifts. Noteworthy in this regard are the $\Delta\delta_C$ in paramagnetic metallocenes which are an order of magnitude higher than typical $\Delta\delta_C$, reflecting the larger range of ^{13}C shifts in these substances.[76]

(4) "The magnitude of the shift is largest where the functional change in mass upon isotopic substitution is largest."[18,23] No exceptions have been noted in ^{13}C nmr. A direct comparison of one-bond effects at ^{13}C can be made between the 0.36 ppm shift per deuterium in propane-2-$\underline{d}_2$ and the 0.023 ppm shift due to ^{16}O/^{18}O substitution in isopropyl alcohol.[20,63]

(5) "The magnitude of the shift is approximately proportional to the number of atoms in the molecule that have been substituted by isotopes."[18,23] Additivity of isotope shifts is observed for one-bond effects, with clear examples from ^{1}H/^{2}H and ^{16}O/^{18}O substitution. There are few reliable data bearing on the

question of additivity for two- and three-bond effects. Additivity was observed in benzene,[2] and approximate additivity for some peptides[107] and carbohydrates.[112] Deviations from additivity were seen for the two-bond effect on [119]Sn in tetramethyltin[39] and on ^{1}H in methane.[16] Nonadditivity in the long-range isotope shifts associated with deuteration of a hyperconjugating methyl group was attributed to an equilibrium isotope effect on conformation.[44]

Few correlations of $\Delta\delta_C$ with structural features can be made, perhaps in part because of the limited number of systematic studies. The $\alpha-\Delta\delta_C$ are affected by substituents at the α-carbon, but the only well established trend is toward larger $\alpha-\Delta\delta_C$ with increased alkyl substitution. Although $\alpha-\Delta\delta_C(^2H)$ vary with ring size in cycloalkanes, there is no clear relation in other compounds to electronic effects, bond strengths, or hybridization. The $\alpha-\Delta\delta_C$ are not affected by steric crowding. For $^{16}O/^{18}O$ substitution, the $\alpha-\Delta\delta_C(^{18}O)$ are larger for single bonds to sp^3 carbons than to sp^2 carbons, but larger in carbonyl groups than in ethers or alcohols. Two-bond isotope shifts, $\beta-\Delta\delta_C$, are reduced in magnitude by nearby electronegative substituents or alkyl branching. In arenes, the magnitude of $\beta-\Delta\delta_C(^2H)$ is related to the C-C bond order. No stereochemical dependence of the $\beta-\Delta\delta_C$ was noted for axial versus equatorial positions in carbohydrates. Downfield $\beta-\Delta\delta_C$ are often observed at electron-deficient carbons for deuterium substitution at hyperconjugating C-H bonds. Reports of three-bond isotope shifts are too few to attempt any correlations.

Isotope effects on coupling constants are quite small, and evidence for the existence of purely intrinsic effects is rather tenuous. Intermolecular interactions have been shown to influence isotope effects on P-H couplings in phosphonates and may play a role in other systems as well. While there is a trend toward slightly reduced coupling constants due to deuterium substitution in organic compounds, experiments with tritiated compounds suggest that part or all of the deuterium effect may be caused by quadrupole relaxation rather than being an intrinsic, mass-related effect.

REFERENCES

1. H. J. Reich, M. Jautelaut, M. T. Messe, F. J. Weigert and J. D. Roberts, J. Am. Chem. Soc., 91 (1969) 7445.

2. H. N. Colli, V. Gold and J. E. Pearson, J. C. S. Chem. Comm., (1973) 408.

3. J. B. Stothers in "Topics in Carbon-13 NMR Spectroscopy," G. C. Levy, ed., Wiley, New York (1974), Vol. 1, Chapter 4.

4. M. J. Shapiro and A. D. Kahle, Org. Magn. Reson., 12 (1979) 235.

5. R. E. Echols and G. C. Levy, J. Org. Chem., 39 (1974) 1321.

6. W. T. Raynes and J. P. Riley, Mol. Phys., 27 (1974) 337.

7. J. M. A. Al-Rawi, J. P. Bloxsidge, C. O'Brien, D. E. Caddy, J. A. Elvidge, J. R. Jones and E. A. Evans, J. C. S. Perkin II, (1974) 1635.

8. L. J. Altman, D. Laungani, G. Gunnarsson, H. Wennerström and S. Forsen, J. Am. Chem. Soc., 100 (1978) 8264.

9. S. I. Chan, L. Lin, D. Clutter and P. Dea, Proc. Natl. Acad. Sci. USA, 65 (1970) 815.

10. G. Gunnarsson, H. Wennerström, W. Egan and S. Forsen, Chem. Phys. Lett., 38 (1976) 96.

11. T. K. Leipert, Org. Magn. Reson., 9 (1977) 157.

12. N. N. Shapetko, Yu. S. Bogacher, I. L. Radushnova and D. N. Shigorin, Dokl. Akad. Nauk. SSSR, 231 (1976) 409.

13. R. H. Martin, J. Moriau and N. Defay, Tetrahedron, 30 (1974) 179.

14. M. Saunders and M. R. Kates, J. Am. Chem. Soc., 99 (1977) 8071.

15. For a leading reference, see M. Saunders, M. R. Kates and G. E. Walker, J. Am. Chem. Soc., 103, (1981) 4623.

16. R. A. Bernheim and B. J. Lavery, J. Chem. Phys., 42 (1965) 1464.

17. M. Alei, Jr. and W. E. Wageman, J. Chem. Phys., 68 (1978) 783.

18. H. Batiz-Hernandez and R. A. Bernheim, Progr. Nucl. Magn. Reson., 3 (1967) 63.

19. F. W. Wehrli and T. Wirthlin, "Interpretation of Carbon-13 NMR Spectra," Heyden, London (1976).

20. J. M. Risley and R. L. Van Etten, J. Am. Chem. Soc., 102 (1980) 4609, and references therein.

21. P. C. Lauterbur, J. Chem. Phys., 42 (1965) 799.

22. W. T. Raynes, A. M. Davies and D. B. Cook, Mol. Phys., 21 (1971) 123.

23. C. J. Jameson, J. Chem. Phys., 66 (1977) 4983.

24. A. D. Buckingham and W. Urland, Chem. Rev., 75 (1975) 113.

25. W. S. Brey, Jr., K. H. Ladner, R. E. Block and W. A. Tallon, J. Magn. Reson., 8 (1972) 406.

26. W. T. Raynes and G. Stanney, J. Magn. Reson., 14 (1974) 378.

27. N. F. Ramsey, Phys. Rev., 87 (1952) 1075.

28. T. W. Marshall, Mol. Phys., 4 (1961) 61.

29. A. Saika and H. Narumi, Can. J. Phys., 42 (1964) 1481.

30. G. E. Maciel, P. D. Ellis and D. C. Hofer, J. Phys. Chem., 71 (1967) 2160.

31. L. S. Bartell, K. Kachitsu and R. J. DeNeui, J. Chem. Phys., 35 (1961) 1211.

32. E. H. Richardson, S. Brodersen, L. Krause and H. L. Welsh, J. Mol. Spectrosc., 8 (1962) 406.

33. A. Loewenstein and M. Shporer, Mol. Phys., 9 (1965) 293.

34. V. M. Mamayev and N. M. Sergeyev, Chem. Phys. Lett., 34 (1975) 317.

35. V. M. Mamayev, N. M. Sergeyev, Kh. A. Orazberdieu, J. Struct. Chem., 18 (1977) 31.

36. R. A. Bernheim and H. Batiz-Hernandez, J. Chem. Phys., 45 (1966) 2261.

37. G. Fraenkel and W. Burlant, J. Chem. Phys., 42 (1965) 3724.

38. M. Murray, J. Magn. Reson., 9 (1973) 326.

39. C. R. Lassigne and E. J. Wells, J. Magn. Reson., 31 (1978) 195.

40. K. L. Servis and F.-F. Shue, J. Am. Chem. Soc., 102 (1980) 7233.

41. J. B. Lambert and L. G. Greifenstein, J. Am. Chem. Soc., 95 (1973) 6150.

42. J. B. Lambert and L. G. Greifenstein, J. Am. Chem. Soc., 96 (1974) 5120.

43. J. W. Timberlake, J. A. Thompson and R. W. Taft, J. Am. Chem. Soc., 93 (1971) 274.

44. D. A. Forsyth, P. Lucas and R. M. Burk, J. Am. Chem. Soc., 104 (1982) 240 , and unpublished observations.

45. V. J. Shiner, Jr., ACS Monogr. No. 167, (1970) Chapter 2.

46. D. E. Sunko and S. Borcic, ACS Monogr. No. 167, (1970) Chapter 3.

47. L. Melander and W. H. Saunders, Jr., "Reaction Rates of Isotopic Molecules," Wiley, New York (1980), Chapter 6.

48. D. J. DeFrees, J. E. Bartmess, J. K. Kim, R. T. McIver, Jr. and W. J. Hehre, J. Am. Chem. Soc., 99 (1977) 6451.

49. D. J. DeFrees, M. Taagepera, B. A. Levi, S. K. Pollack, K. D. Summerhays, R. W. Taft, M. Wolfsberg and W. J. Hehre, J. Am. Chem. Soc., 101 (1979) 5532.

50. A. P. Tulloch, Can. J. Chem., 55 (1977) 1135.

51. R. L. Elsenbaumer and H. S. Mosher, J. Org. Chem., 44 (1979) 600.

52. A. G. McInnes, J. A. Walter and J. L. C. Wright, Tetrahedron Lett., (1979) 3245.

53. D. G. Morris and A. M. Murray, J. C. S. Perkins II, (1976) 1579.

54. L. M. Stephenson, R. V. Gemmer and S. P. Current, J. Org. Chem., 42 (1977) 212.

55. D. Lauer, E. L. Motell, D. D. Traficante and G. E. Maciel, J. Am. Chem. Soc., 94 (1972) 5535.

56. Yu. K. Grishin, N. M. Sergeyev, Yu. A. Ustynyuk, Mol. Phys., 22 (1971) 711.

57. E. Breitmaier, G. Jung, W. Voelter and L. Pohl, Tetrahedron 29 (1973) 2485.

58. G. C. Levy and J. D. Cargioli, J. Magn. Reson., 6 (1972) 143.

59. D. Doddrell and I. Burfitt, Austral. J. Chem., 25 (1972) 2239.

60. A. P. Tulloch and M. Mazurek, J. C. S. Chem. Comm., (1973) 692.

61. J. M. A. Al-Rawi, J. A. Elvidge and J. R. Jones, J. C. S. Perkin II, (1975) 449.

62. F. W. Wehrli, D. Jeremic, M. L. Mihailovic and S. Milosavljevic, J. C. S. Chem. Comm., (1978) 302.

63. R. E. Wasylishen and T. Schaeffer, Can. J. Chem., 52 (1974) 3247.

64. L. A. Paquette, C. W. Doecke, F. R. Kearney, A. F. Drake and S. F. Mason, J. Am. Chem. Soc., 102 (1981) 7228.

65. A. A. Borisenko, N. M. Sergeyev and Yu. A. Ustynyuk, Mol. Phys., 22 (1971) 715.

66. J. B. Stothers, "Carbon-13 NMR Spectroscopy," Academic Press, New York (1972).

67. D. C. McKean, J. L. Duncan and L. Batt, Spectrochim. Acta, 29A (1973) 1037.

68. R. Aydin and H. Günther, J. Am. Chem. Soc., 103 (1981) 1301.

69. K. J. Friesen and R. E. Wasylishen, J. Magn. Reson., 41 (1980) 189.

70. M. Saunders, M. R. Kates, K. B. Wiberg and W. Pratt, J. Am. Chem. Soc., 99 (1979) 8072.

71. R. A. Bell, C. L. Chan and B. G. Sayer, J. C. S. Chem. Comm., (1972) 67.

72. H. Günther, H. Seel and M.-E. Günther, Org. Magn. Reson., 11 (1978) 97.

73. W. Kitching, M. Bullpitt, D. Doddrell and W. Adcock, Org. Magn. Reson., 6 (1974) 289.

74. G. W. Buchanan and R. S. Ozubko, Can. J. Chem., 53 (1975) 1829.

75. G. L. Lebel, J. D. Laposa, B. G. Sayer and R. A. Bell, Analyt. Chem., 43 (1971) 1500.

76. F. H. Köhler and W. Prössdorf, J. Am. Chem. Soc., 100 (1978) 5970.

77. F. A. Bovey, K. B. Abbas, F. C. Schilling and W. H. Starnes, Macromolec., 8 (1975) 437.

78. M. Saunders, L. Telkowski and M. R. Kates, J. Am. Chem. Soc., 99 (1977) 8070.

79. J. B. Stothers, C. T. Tan, A. Nickon, F. Huang, R. Sridhar and R. Weglein, J. Am. Chem. Soc., 94 (1972) 8581.

80. D. H. Hunter, A. L. Johnson, J. B. Stothers, A. Nickon, J. L. Lambert and D. F. Covey, J. Am. Chem. Soc., 94 (1972) 8582.

81. D. H. Hunter and J. B. Stothers, Can. J. Chem., 51 (1973) 2884.

82. J. M. Risley and R. L. Van Etten, J. Am. Chem. Soc., 101 (1979) 252.

83. D. J. Darensbourg and B. J. Baldwin, J. Am. Chem. Soc., 101 (1979) 6447.

84. D. J. Darensbourg, J. Organomet. Chem., 174 (1979) C70-C76.

85. J. C. Vederas, J. Am. Chem. Soc., 102 (1980) 374.

86. J. M. Risley and R. L. Van Etten, J. Am. Chem. Soc., 102 (1980) 6699.

87. P. S. Nielsen, R.S. Hansen and H. J. Jakobsen, J. Organomet. Chem., 114 (1976) 145.

88. F. J. Weigert and J. D. Roberts, J. Am. Chem. Soc., 94 (1972) 6021.

89. J. Jokisaari, Org. Magn. Reson., 11 (1978) 157.

90. K. Frei and H. J. Bernstein, J. Chem. Phys., 38 (1963) 1216.

91. R. E. Wasylishen, D. H. Muldrew and K. J. Friesen, J. Magn. Reson., 41 (1980) 341.

92. W. Buchner and D. Schentzow, Org. Magn. Reson., 7 (1975) 615.

93. S. A. Linde and H. J. Jakobsen, J. Magn. Reson., 17 (1975) 411.

94. P. A. J. Gorin, Can. J. Chem., 52 (1974) 458.

95. P. A. J. Gorin and M. Mazurek, Can. J. Chem., 53 (1975) 1212.

96. F. Balza, N. Cyr, G. K. Hamer, A. S. Perlin, H. J. Koch and R. S. Stuart, Carbohydr. Res., 59 (1977) C7-C11.

97. F. J. Weigert and J. D. Roberts, J. Am. Chem. Soc., 89 (1967) 2967.

98. R. Biehl, K. Hinrichs, H. Kurreck, W. Lubitz, U. Mennenga and K. Roth, J. Am. Chem. Soc., 99 (1977) 4278.

99. T. Bundgaard, H. J. Jakobsen and E. J. Rahkamaa, J. Magn. Reson., 19 (1975) 345.

100. J. B. Stothers, C. T. Tan and K. C. Teo, Can. J. Chem., 31 (1973) 2893.

101. F. A. L. Anet and A. H. Dekmezian, J. Am. Chem. Soc., 101 (1979) 5449.

102. F. A. L. Anet, A. K. Cheng and J. Krane, J. Am. Chem. Soc., 95 (1973) 7877.

103. F. A. L. Anet, V. J. Basus, A. P. W. Hewett and M. Saunders, J. Am. Chem. Soc., 102 (1980) 3945.

104. N. Gurudata, Can. J. Chem., 50 (1972) 1956.

105. J. Feeney, P. Partington and G. C. K. Roberts, J. Magn. Reson., 13 (1974) 268.

106. R. A. Newmark and J. R. Hill, J. Magn. Reson., 21 (1976) 1.

107. H. K. Ladner, J. J. Led and D. M. Grant, J. Magn. Reson., 20 (1975) 530.

108. J. J. Led and S. B. Petersen, J. Magn. Reson., 33 (1979) 603.

109. I. Kurobane and L. C. Vining, Tetrahedron Lett., (1978)
4633.

110. D. Gagnaire and M. Vincendon, J. C. S. Chem. Comm., (1977)
509.

111. S.-C. Ho, H. J. Koch and R. S. Stuart, Carbohydr. Res., 64
(1978) 251.

112. P. E. Pfeffer, K. M. Valentine and F. W. Parrish, J. Am.
Chem. Soc., 101 (1979) 1265.

113. Y. Kanazawa, J. D. Baldeschwieler and N. C. Craig, J. Mol.
Spectrosc., 16 (1965) 325.

114. G. Fraenkel, Y. Asahi, H. Batiz-Hernandez and R. A. Bernheim, J. Chem. Phys., 44 (1966) 4647.

115. A. L. Allred and W. D. Wilk, J. C. S. Chem. Comm., (1969)
273.

116. G. B. Savitsky, L. G. Robinson, W. A. Tallon and L. R.
Womble, J. Magn. Reson., 1 (1969) 139.

117. G. M. Ford, L. G. Robinson and G. B. Savitsky, J. Magn.
Reson., 4 (1970) 109.

118. G. V. Smith, W. A. Boyd and C. C. Hinckley, J. Am. Chem.
Soc., 93 (1971) 6319.

119. C. H. DePuy, P. C. Fünfschilling and J. M. Olson, J. Am.
Chem. Soc., 98 (1976) 276.

120. D. D. Traficante and G. E. Maciel, J. Am. Chem. Soc., 87
(1965) 4917.

121. W. R. Young and C. S. Yannoni, J. Am. Chem. Soc., 91 (1969)
4581.

122. G. A. Olah, P. W. Westerman and D. A. Forsyth, J. Am. Chem.
Soc., 97 (1975) 3419, and references therein.

123. R. J. Spear, D. A. Forsyth and G. A. Olah, J. Am. Chem.
Soc., 98 (1976) 2493.

124. G. Angelini, M. Speranza, A. L. Segre and L. J. Altman, J.
Org. Chem., 45 (1980) 3291.

125. M. H. Pendlebury and L. Phillips, Org. Magn. Reson., 4
(1972) 529.

126. P. J. Mitchell and L. Phillips, J. C. S. Chem. Comm., (1975)
908.

127. J. A. Pople, W. G. Schneider and H. J. Bernstein, "High-
Resolution Nuclear Magnetic Resonance," McGraw-Hill, New
York (1959), p. 188.

128. C. Brevard, J. P. Kintzinger and J. M. Lehn, J. C. S. Chem. Comm., (1969) 1193.

129. N. Muller and R. H. Birkhahn, J. Chem. Phys., 43 (1965) 4540.

130. G. Fraenkel and W. J. Burlant, J. Chem. Phys., 43 (1965) 4541.

131. W. M. Litchman, M. Alei, Jr., and A. E. Florin, J. Chem. Phys., 50 (1969) 1897.

132. R. A. Bernheim and H. Batiz-Hernandez, J. Chem. Phys., 40 (1964) 3446.

133. a) W. J. Stec, N. Goddard and J. R. Van Wazer, J. Phys. Chem., 75 (1971) 3547. b) W. McFarlane and D. S. Rycroft, Mol. Phys., 24 (1972) 893.

134. A. K. Jameson and C. J. Jameson, J. Magn. Reson., 32 (1978) 455.

135. P. Diehl and T. Leipert, Helv. Chim. Acta, 47 (1964) 545.

136. B. E. Smith, B. D. James and R. M. Peachey, Inorg. Chem., 16 (1977) 2057.

137. C. Schumann and H. Dreeskamp, J. Magn. Reson., 3 (1970) 204.

138. G. Pfisterer and H. Dreeskamp, Ber. Bunsenges. Phys. Chem., 73 (1969) 654.

139. E. A. V. Ebsworth and J. J. Turner, J. Chem. Phys., 36 (1962) 2628.

140. N. M. Sergeyev and V. N. Solkan, J. C. S. Chem. Comm. (1975) 12.

141. N. Muller, J. Chem. Phys., 42 (1965) 4309.

142. R. R. Fraser, M. A. Petit and M. Miskow, J. Am. Chem. Soc., 94 (1972) 3253.

143. J. P. Bloxsidge, J. A. Elvidge, J. R. Jones, R. B. Mane and M. Saljoughian, Org. Magn. Reson., 12 (1979) 574.

Chapter 2

THE EFFECT OF PRESSURE ON KINETIC ISOTOPE EFFECTS

NEIL S. ISAACS

Chemistry Department, University of Reading, Reading, England.

CONTENTS

I. EFFECTS OF PRESSURE ON THE RATES OF LIQUID PHASE REACTIONS

The application of pressure to reactions in the liquid state has long been known to be capable of bringing about a change in rate.[1-6] Unlike a rise in temperature which always causes an increase in rate, hydrostatic pressure may bring about either a rate acceleration or a retardation as shown in Fig. 1, the extent

of which vary roughly between -0.08 to +0.3% per bar.[*] In order
to measure these small perturbations, kinetic studies need to be
extended into the kbar range. This poses no insurmountable
problems however, and such measurements are made routinely in many
parts of the world.

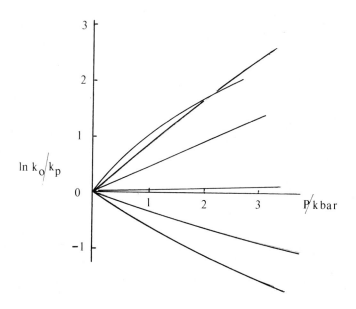

Fig. 1. Typical rate-pressure plots: (1) dimerization of cyclo-
pentadiene at 20°; (2) nitration of benzene, 0°; (3) nitration of
toluene, 0°; (4) hydrolysis of s-trioxan; (5) rearrangement of
N-chloroacetanilide, 25°; (6) hydrolysis of $Co(NH_3)_5OSO_3^+$ by OH^-.

Pressure and volume are conjugate variables as expressed in
Boyle's Gas Law, pV = const. Chemical equilibrium in solution
responds to pressure according to the relationship shown in eq. 1:

$$\frac{-RT \; \partial \ln K}{\partial p} = \Delta \bar{V} \tag{1}$$

where K is the equilibrium constant (expressed in pressure-

* 1 bar = 0.98692 standard atmospheres = $10^5 Pa = 10^5 \; N \; m^{-2}$;
  1 kbar = 1,000 bar

independent units) and $\Delta\overline{V}$ is the change in partial molar volumes of products over reagents, i.e. the volume change per mole for the reaction. The same formal treatment may be applied to rate processes which are considered to proceed by way of an equilibrium between reagents and transition state, thus (eq. 2):

$$-\frac{RT}{}\frac{\partial \ln k}{\partial p} = \Delta V^{\neq} = (\overline{V}_{reagents} - \overline{V}^{\neq}) \tag{2}$$

$\Delta V^{\neq}$ is known as the volume of activation and is the difference in partial molar volumes between the reagents on the one hand and that of the transition state, $\overline{V}^{\neq}$, on the other. It will be observed that a negative volume of activation implies a rate which is accelerated by pressure and conversely. Partial molar volumes of solutes are readily measured from the densities of dilute solutions,[7] (eq.3):

$$\overline{V}_{c \to o} = \frac{1000(\rho_o - \rho)}{\rho_o c} + \frac{M}{\rho_o} \tag{3}$$

where ρ, ρ_o are densities of solution and solvent respectively. It is, therefore, possible by the use of eq. 2 to measure the volume of the transition state, a quantity which is of value in mechanistic investigations.

As shown by Fig. 1, a plot of ln k against pressure may not be linear, tending to curve towards the pressure axis. This indicates that the volume of activation diminishes with pressure and may be attributed to the difference in compressibilities between the reagents on the one hand and the transition state on the other. This quantity is defined as the second pressure derivative, $-RT\,\partial^2 \ln k/\partial p^2 = \Delta\beta^{\neq}$, known as the compressibility coefficient of activation. If the rate data is sufficiently precise, $\Delta\beta^{\neq}$ may be evaluated and has some value in mechanistic interpretation especially in that of inorganic reactions.[8] It is conventional to express the volume of activation at 1 bar, properly symbolised $\Delta V_o^{\neq}$, which will in further discussion be assumed. No analytical expression is derivable from theory by which $\Delta V_o^{\neq}$ may be calculated. It is usual to fit the rate data to a quadratic or higher order polynomial by a least squares procedure, (eq.4):

$$\ln k = A + Bp + Cp^2 + \ldots \tag{4}$$

when $\Delta V_0^{\neq} = -BRT$, $\Delta\beta^{\neq} = 2RTC$. Other methods of handling the data are discussed by Kelm[9] and Isaacs[6].

II. THE MEASUREMENT OF RATES AT HIGH PRESSURE

Adaptations of all the usual kinetics techniques have been devised for use at high pressure. Measurements are usually made at pressures ranging up to 1 kbar and often beyond this, but for the evaluation of $\Delta V_0^{\neq}$ by extrapolation it is important to concentrate on the low pressure end of this range. The reacting solution may be sealed into a steel vessel which may be raised to the desired pressure by a suitable pump or intensifier supplying oil. Provision must be made to isolate the sample from the pressurising fluid and also from the steel of the vessel which would affect many reactions. An inner vessel of P.T.F.E.* is commonly adopted, sealed by a piston or mercury reservoir so that pressure is transmitted to the reactants. Two such devices are illustrated in Fig. 2; the first permits samples to be withdrawn through a needle valve for conventional analysis. The second contains a conductivity cell by means of which appropriate reactions may be monitored in situ.

Windows of quartz or sapphire may be fitted to a pressure vessel enabling analysis by spectrophotometry in the ultraviolet or visible to be used; such equipment is commercially available.[12] Several research groups have now designed and built high pressure NMR probes which will fit into conventional spectrometers, Fig. 3 for example.[13]

Considerable time is required to seal solutions into a pressure vessel, and to allow temperature equilibration after pressurisation which means that these techniques are suitable only for fairly slow reactions. Apparatus has been constructed to permit the mixing of two solutions after pressurisation[14,15] which uses externally operated electromagnets. In an optical cell, devices of this type could permit the study of bimolecular reactions with half-lives of several minutes. Far more versatile are the various designs of stopped-flow spectrometer which have now been published, for example Fig. 4A, which permit a series of kinetic runs to be made

* poly(tetrafluoroethene)

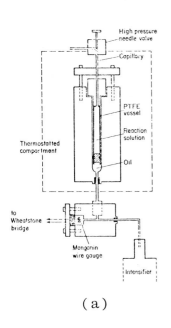

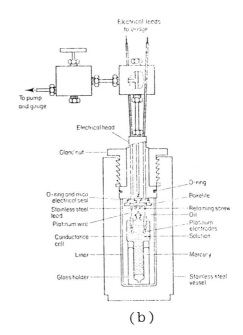

(a)   (b)

Fig. 2. a. A sampling vessel for use under pressure,[10]
         b. conductivity cell for use under pressure.[11]

for each loading while pressure may be varied and can achieve a
time resolution of milliseconds.[16]   Very fast reactions in
solution can only be studied by perturbation of an equilibrium
using T-jump or p-jump techniques. Both of these principles
have been applied to high pressure studies, Fig. 4B[17,18,19], and
some results are discussed below. Radical reactions which may be
followed by e.s.r. can be conducted in a thick-walled glass tube,
pressurised directly;[20]   carefully prepared tubes will withstand
several hundred bar.

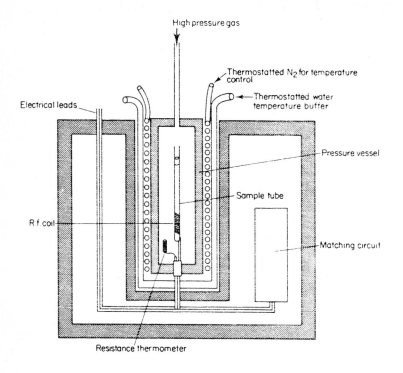

High pressure gas

Thermostatted N₂ for temperature control

Thermostatted water temperature buffer

Electrical leads

Pressure vessel

Sample tube

R.f. coil

Matching circuit

Resistance thermometer

Fig. 3:   Nmr probe for use under high pressure.[13]

III. MECHANISTIC FEATURES WHICH CONTRIBUTE TO THE ACTIVATION VOLUME

Some 2000 volumes of activation have been reported [3,21] relating to reactions of almost all types, both organic and inorganic. The interpretation of these values has contributed much to the understanding of reaction mechanisms. The prediction of the volume of activation for a reaction according to a particular mechanistic scheme must take into account several contributing factors. A.Bond-formation and -fission. The association of two molecules in the formation of a bimolecular transition state is associated in general with a reduction in volume and in consequence an increase in rate with pressure. This is illustrated by examples shown in Table 1. Atom or group transfer reactions in which no change in charge occurs include radical hydrogen abstraction, (ex. 3), neutral S_N2 processes (ex. 7) and ketone reductions by borohydride (ex.11). For this type of activation process, namely

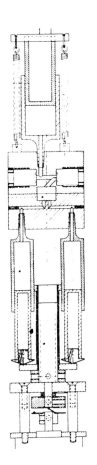
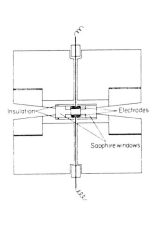

Fig. 4: A, Stopped-flow system designed for high pressures

  B, Temperature-jump apparatus for high pressures

$X + Y-Z \rightarrow [X...Y...Z] \rightarrow$ products

$\Delta V^{\neq}$ usually lies in the range -5 to -15 $cm^3 mol^{-1}$. Neutral additions such as the polymerization[25] (ex. 2) and aromatic electrophilic substitution (ex. 10) are somewhat more negative, -15 to -25 cm^3 mol^{-1}, since there is no compensating bond-cleavage component.

$X + Y \rightarrow [X...Y] \rightarrow$ products

The simultaneous formation of two bonds as in the Diels-Alder reaction (ex. 4) is accompanied by a volume reduction in the range -30 to -35 $cm^3 mol^{-1}$, roughly double that for the formation of one covalent bond and the values are very close to the overall volume

Table 1

Volumes of Activation for Some Typical Reactions

	$\Delta V_o^{\neq}$ / cm^{3}mol^{-1}	ref.
1. t-BuO–OtBu $\xrightarrow[C_7H_{16}]{\Delta}$ [tBuO...OtBu] → 2 tBuO·	+10	24
2. (see scheme)		
3. Ph$_2$N–N·$_{pic}$ H–OAr → [Ph$_2$N–N...H..OAr]$_{pic}$ → Ph$_2$N–NH$_{pic}$ ·OAr	–13	26
4. (see scheme)	–30	27
5. (see scheme)	–55	28

Row 2 scheme:

$$\begin{matrix} \text{Ph} \\ |\\ -\text{CH}\cdot \\ |\\ -\text{CH}_2 \end{matrix} \quad \begin{matrix}\text{Ph}\\|\\-\text{CH}\to\\|\\ \text{CH}_2 \end{matrix} \quad\to\quad \left[\begin{matrix} \text{Ph}\\|\\-\text{CH}\\|\\-\text{CH}_2 \end{matrix} \quad \begin{matrix}\text{Ph}\\|\\ \text{CH}\\ \vdots\\-\text{CH}_2 \end{matrix}\right]^{\cdot} \quad\to\quad \begin{matrix}\text{Ph}\\|\\-\text{CH}\\|\\-\text{CH}_2 \end{matrix}\quad\begin{matrix}\text{Ph}\\|\\ \text{CH}\cdot\\|\\ \text{CH}_2 \end{matrix}$$

Row 4 scheme:

$$\text{Me}\!\!\diagup\!\!\diagdown \quad + \quad \overset{\text{COOMe}}{\underset{\diagdown}{}} \quad\to\quad \left[\begin{matrix}\text{Me}\\ \diagup\diagdown \\ \text{COOMe}\end{matrix}\right] \quad\to\quad \begin{matrix}\text{Me}\\ \diagup\diagdown \\ \text{COOMe}\end{matrix}$$

Row 5 scheme:

$$\begin{matrix}\text{NC}\\ \text{NC}\end{matrix}\!\!\diagdown\!\!\diagup\!\!\begin{matrix}\text{CN}\\ \text{CN}\end{matrix} \quad + \quad \diagup\diagdown\!\!-\text{OEt} \quad\to\quad \left[\begin{matrix}\text{NC} \quad \overset{+}{|}\text{OEt}\\ \text{NC} \quad \underset{|}{} \\ \text{NC} \quad \text{CN}\end{matrix}\right] \quad\to\quad \begin{matrix}\text{NC}\quad\square\quad\text{OEt}\\ \text{NC}\qquad\text{CN}\end{matrix}$$

		$\Delta V_o^{\neq}/\text{cm}^3\text{mol}^{-1}$	ref.
6.	$\begin{array}{c}\text{Ph}\\\text{C-Cl}\\\text{Me}\quad\text{Me}\end{array} \rightarrow \left[\begin{array}{c}\text{Ph}\\\text{C}^{\delta+}\cdots\text{Cl}^{\delta-}\\\text{Me}\quad\text{Me}\end{array}\right] \rightarrow \begin{array}{c}\text{Ph}\\\text{C-OH}\\\text{Me}\quad\text{Me}\end{array}$	-12	29
7.	$\begin{array}{c}\text{H}\\\text{I}^-\;\text{C-Cl}\\\text{H}\quad\text{Et}\end{array} \rightarrow \left[\begin{array}{c}\delta-\;\text{I}\cdots\begin{array}{c}\text{H}\\\text{C}\\\text{H}\quad\text{Et}\end{array}\cdots\text{Cl}^{\delta-}\end{array}\right] \rightarrow \begin{array}{c}\text{H}\;\text{H}\\\text{I-C}\\\text{Et}\end{array}$	-6	30
8.	$\begin{array}{c}\text{H}\;\text{H}\\\text{Et}_3\text{N:}\;\;\text{C-I}\\\text{Me}\end{array} \rightarrow \left[\begin{array}{c}\text{Et}_3\text{N}^{\delta+}\cdots\begin{array}{c}\text{H}\;\text{H}\\\text{C}\\\text{Me}\end{array}\cdots\text{I}^{\delta-}\end{array}\right] \rightarrow \text{Et}_3\overset{+}{\text{N}}\text{-CH}_2\text{Me}\;\text{I}^-$	$\begin{array}{c}\text{(MeOH)}\;\;-22\\(\text{C}_6\text{H}_6)\;\;-50\end{array}$	$\begin{array}{c}31\\32\end{array}$
9.	$\begin{array}{c}\text{H}\;\text{H}\\\text{PhO}^-\;\;\text{C-}\overset{+}{\text{S}}\text{Et}_2\\\text{Me}\end{array} \rightarrow \left[\begin{array}{c}\delta-\\\text{PhO}\cdots\begin{array}{c}\text{H}\;\text{H}\\\text{C}\\\text{Me}\end{array}\cdots\overset{\delta+}{\text{SEt}_2}\end{array}\right] \rightarrow \text{PhOCH}_2\text{Me} + \text{SEt}_2$	$+12$	33
10.	$\text{PhH} \xrightarrow{\text{HNO}_3} \text{Ph}\overset{\overset{+}{\text{H}}}{\diagdown}\text{NO}_2 \longrightarrow \text{PhNO}_2$	-22	34

	$\Delta V_0^{\neq}\,/\,\mathrm{cm^3\,mol^{-1}}$	ref.

11.

$$\mathrm{Ph}{\overset{}{\underset{\mathrm{Me}}{}}}\mathrm{C{=}O} + \mathrm{BH_4^-} \;\rightarrow\; \left[\mathrm{Ph}{\overset{}{\underset{\mathrm{Me}}{}}}\mathrm{C{\cdots}O\cdots H\cdots BH_3} \right]^- \;\rightarrow\; \mathrm{Ph}{\overset{}{\underset{\mathrm{Me}}{}}}\mathrm{C{-}O{-}(B} \backslash\!\!/\,)\quad\overset{}{\underset{\mathrm{H}}{}}$$

−11   35

12.

$$\mathrm{Ph{-}\overset{O}{C}{-}CH_2{-}\overset{O}{C}{-}O^-} + \mathrm{H_2O} \;\rightarrow\; \left[\mathrm{Ph{-}\overset{\cdots O}{C}{\cdots}CH_2{\cdots}CO_2} \right]^- \;\rightarrow\; \mathrm{Ph{-}\overset{O}{C}{-}CH_3} + \mathrm{OH^-} + \mathrm{CO_2}$$

+5   36

13.

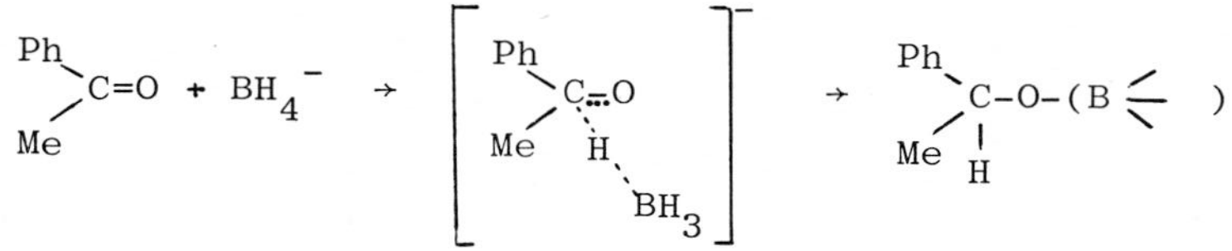

$$\mathrm{Ph{-}C}{\overset{O}{\underset{O{-}H}{}}}\quad \mathrm{H{-}O}{\overset{}{\underset{O}{}}}{C{-}Ph} \;\rightleftharpoons\; \mathrm{Ph{-}C}{\overset{O{\cdots}H{\cdots}O}{\underset{O{\cdots}H{\cdots}O}{}}}{C{-}Ph}$$

$(\Delta\bar V)$

−13   6

14.

Cl–(2,4-dinitrophenyl) + tBuNH₂ → [Meisenheimer-type intermediate: Cl, NH₂tBu, 2-NO₂, 4-NO₂] → (2,4-dinitrophenyl)NHtBu

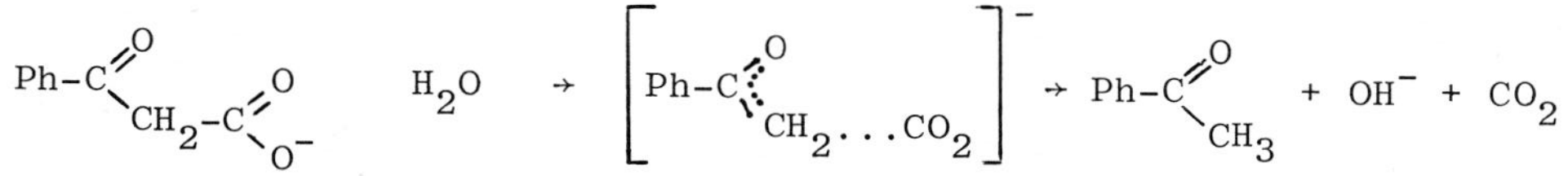

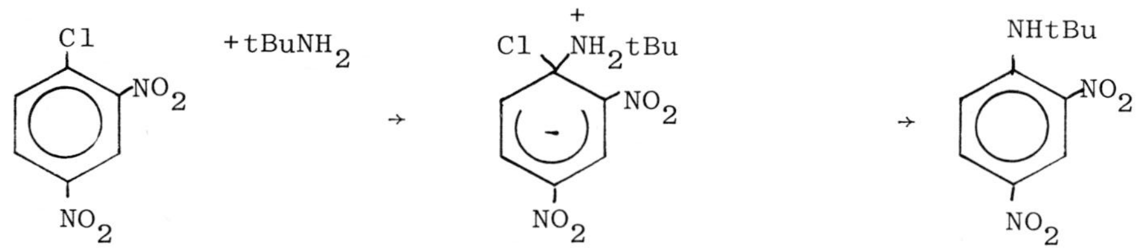

−35   6

change for the reaction. Hydrogen-bond formation causes a volume reduction; -13 cm^3mol^{-1} for formation of the two hydrogen bonds in a carboxylic acid dimer, (ex. 13).

Conversely, bond fission is associated with an increase in volume although the magnitudes of the volume changes seem always to be less than those for bond formation. This probably means that the transition state is reached with a quite small lengthening of the

$$X-Y \rightarrow [X...Y] \rightarrow products$$

bond which is being broken. Examples include the homolysis of peroxides, (ex. 1) and the fragmentation of the β-ketoacid, (ex. 12). Reactions of these types are retarded by the application of pressure.

B. Changes in solvation

The experimental value of $\Delta V_o^{\neq}$ may be dominated by changes in the solvation of the reagents as they pass to the transition state. In general, a strong electric field gradient, such as that associated with a charge or dipole, brings about close packing or compression of solvent molecules in its vicinity, a process known as electrostriction. The volume change ΔV_e, known as the volume of electrostriction, which is brought about by a spherical ion, radius r and charge q in a continuous medium of dielectric constant ε, is derivable from electrostatic theory and given by the Drude-Nernst equation (eq. 5):

$$\Delta V_e = - \frac{q^2}{2r} \cdot \frac{\partial (1/\varepsilon)}{\partial p} = - \frac{q^2 \phi}{2r} \tag{5}$$

The model is a crude one, though nothing else is readily available, but it serves to explain the effect of charge and dielectric constant on values of $\Delta V_o^{\neq}$. The q^2 term indicates that the accumulation of multiple charges upon an ion will have increasingly large electro-strictive effects; this is illustrated, Table 2, by values of partial molar volumes for some ions of comparable Van der Waals radius. Indeed, ions with a high charge/volume ratio may have negative partial molar volumes. It follows that a reaction in which a dipolar transition state is formed from neutral reagents will be accompanied by a large electrostrictive effect and a negative contribution to $\Delta V^{\neq}$, whereas the relaxation of solvent

TABLE 2

Partial Molar Volumes of Some Ions and Non-electrolytes[7]
$\overline{V}/cm^3 mol^{-1}$ in water at 25° unless otherwise stated.

$\overline{V}$			$\overline{V}$
H^+ =	0.0 (convention)*		
OH^-	-4.04	ClO_4^-	44.12
Li^+	-0.88	MnO_4^-	42.5
Na^+	-1.21	SO_4^{2-}	13.98
K^+	9.02	CrO_4^{2-}	19.7
Cs^+	21.34	AsO_4^{3-}	-15.6
F^-	-1.16	Ph_4B^- (in MeOH)	259
Cl^-	17.83	Ph_4C	295
Br^-	24.71		
I^-	36.22		

* The best value for $\overline{V}$ (H^+) is -4 $cm^3 mol^{-1}$

when charges become neutralised may result in a volume increase.
The second point which follows from the Drude-Nernst equation is
the importance of dielectric constant or rather the quantity
$\Phi = \partial(1/\varepsilon)/\partial p$. This roughly follows the reverse of an order of
solvent 'polarity' as measured, for example, by solvatochromic dyes –
the E_T scale.   The volume of electrostriction is, therefore,
greatest for solvents of low 'polarity' since in these media the
electric field in the vicinity of the ion is experienced at a
greater distance than is the case with a high ε and, in addition,
such solvents are usually more highly compressible than are polar
solvents. As a result of this effect it is found that a dipolar
S_N2 reaction - the Menshutkin reaction, Table 1, ex. 8 - is
accelerated by pressure to a far greater extent than the neutral
counterpart, (ex. 7). $\Delta V_o^{\neq}$ is very solvent-sensitive and more
negative in the less polar solvents. The (2+2) cycloaddition (ex.5)
has been shown to occur by a 2-step process by way of a dipolar
intermediate.[22]   This causes an even more negative activation volume
than that for the concerted Diels-Alder reaction. The hydrolysis of
cumyl chloride (ex. 6) is well established as occurring by carbon-
chlorine bond heterolysis.[23] Far from showing a positive volume of

activation, $\Delta V^{\neq}_o$ is similar to that for an S_N2 process due to the electrostrictive influence of the developing ionic centres.

Charge neutralisation should have the reverse effect and bring about relaxation of electrostatically bound solvent with consequent increase in volume. This is clearly seen in the S_N2 displacement (ex. 9) between oppositely charged reagents which is strongly retarded by pressure although it is an associative process. In general one can only expect a moderate correlation between $\Delta V^{\neq}$ and Φ since solvation forces other than electrostatic types are usually present. In particular donor-acceptor interactions between solvent and solute (specific solvation) may intrude and cause large deviations from predicted orders, Fig. 5.

These two principles in the main determine the magnitude of $\Delta V^{\neq}$. In the analysis of an elementary reaction one process only needs to be considered. A multi-step reaction, however, requires consideration of all reactions up to the formation of the rate-determining transition state. For example, the following situation is frequently observed:

$$A + B \underset{}{\overset{K}{\rightleftharpoons}} C$$

$$C \xrightarrow[\text{slow}]{k_1} \text{products}$$

and $\Delta V^{\neq}_{obs} = \Delta \overline{V} + \Delta V^{\neq}_1$ where $\Delta \overline{V}$ is the volume change for the pre-equilibrium. The ubiquity of complex reactions is often a barrier to the interpretation of pressure effects since the two components cannot often be disentangled. It is frequently assumed, for instance, (for want of definite information) that a pre-equilibrium proton transfer characteristic of many acid-catalysed reactions has $\Delta \overline{V} \sim 0$ when only oxygen bases are involved. This is not necessarily the case for a proton transfer between two dissimilar bases as the following example will show. The hydrolysis of chloroform has $\Delta V^{\neq}_{obs} = +16$ $cm^3 mol^{-1}$ and is believed to occur by the two-step sequence.

$$CHCl_3 + OH^- \underset{k_{-1}}{\overset{k_1}{\rightleftharpoons}} CCl_3^- + H_2O \qquad \Delta \overline{V} = +9 \ cm^3 mol^{-1}$$

$$CCl_3^- \xrightarrow{k_2} :CCl_2 + Cl^- \qquad \Delta V^{\neq} = \dfrac{+7 \ cm^3 mol^{-1}}{+16 \ cm^3 mol^{-1}}$$

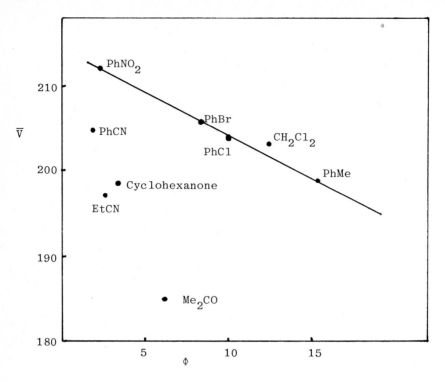

Fig. 5. Volume of the transition state for the Menshutkin Reaction

$$PhCOCH_2Br + N\bigcirc \longrightarrow PhCOCH_2\overset{+}{N}\bigcirc \quad Br^-$$

as a function of the solvent parameter, Φ.[72]

Here, $\Delta \bar{V}$ can be determined by a deuterium exchange experiment
and $\Delta V_2^{\neq}$ is found to be positive in accordance with the rate-
determining dissociation to carbene and chloride ion. Probably this
last figure contains a negative component from the increased electrostriction expected since the charge is transferred from a large
delocalised ion to a small charge-localised one.

IV. EFFECTS OF PRESSURE ON VIBRATIONAL FREQUENCIES

Although kinetic isotope effects are usually classified as
primary, secondary or solvent effects, such distinctions are
convenient rather than reflecting real differences in their origins.
The ultimate origin of all types of isotope effects lies in changes in
vibrational frequencies brought about by isotopic substitution. The
Bigeleisen equation (eq. 6) expresses this fundamental relationship.

$$
\frac{k_H}{k_D} = \frac{\nu_H^{\neq}}{\nu_D^{\neq}} \left\{ \frac{\prod (\nu_i^H/\nu_i^D)\, e^{-\Delta u_i/2} \cdot \left(\frac{1-e^{-u_i^D}}{1-e^{-u_i^H}}\right)}{\prod (\nu_j^H/\nu_j^D)\, e^{-\Delta u_j/2} \cdot \left(\frac{1-e^{-u_j^D}}{1-e^{-u_j^H}}\right)} \right\}
\tag{6}
$$

where u = hν/kT and the ν's are vibrational frequencies for all modes
of the two isotopically distinguishable molecules, including the
'imaginary' vibration, $\nu^{\neq}$, constituting motion of the hydrogen along
the reaction coordinate. We must therefore first enquire whether
pressure affects vibrational frequencies. Both infrared and Raman
spectra have been measured under high pressure conditions. Using a
diamond anvil cell, indeed spectra may be obtained of samples com-
pressed above 500 kbar.[36,37] The much lower pressures used in
kinetic studies are not capable of affecting force constants or
covalent bond lengths although by affecting average intramolecular
distances small shifts in medium-dependent vibrations are found.
The hydrogen-bonded -O-D stretch in aqueous solution at about 2100 cm^{-1}
decreases by about 2.3 $cm^{-1}kbar^{-1}$.[38] The first harmonic of the
-OH stretch in some phenols, at around 7000 cm^{-1}, increases by
6-14 $cm^{-1}kbar^{-1}$.[26] Crystal vibrational modes may be affected, for
example, the fundamental of diamond by 0.28 $cm^{-1}kbar^{-1}$. Some values
from the literature are shown in Table 3. It is true to say, however,
that while pressure effects on vibrational frequencies are

TABLE 3

Pressure-induced infrared shifts.[6,26]

			$\Delta \bar{\nu}$ (cm^{-1} kbar^{-1})		
Vibration	Solvent		None	CS$_2$	CFCl$_3$
C-H stretch					
CHCl$_3$				-1.72	+0.19
CH$_2$Cl$_2$	Symm			-0.48	+0.25
	Asymm			+0.20	+0.92
CHI$_3$				-1.22	-
C-F stretch; CFCl$_3$			-1.77		
O-H stretch; BuO-H				-1.1	
PhO-H					-1.46
ArO-H,	2nd harmonic			-6 to -14	
HO-D	(in H$_2$O)		-2.3		
N-H stretch; PhNH$_2$,	Symm				-0.08
	Asymm				0
S-H stretch; nPrS-H				-0.56	

Values are averages over 6-10 kbar and are not linear.

unpredictable, they are small. In particular, at pressures in the
range 1-3 kbar at which kinetic studies are made, changes in the
isotopic ratios of frequencies, ν_H/ν_D, by pressure would be negligible.
The effect of pressure upon the 'imaginary' frequencies, ν',
associated with the motion of the hydrogen atom between two nuclei,
is harder to assess. Since the transfer of hydrogen between two
much larger masses is the same type of motion as the asymmetric
stretch of a triatomic system we can, however, replace the ratio
ν_H'/ν_D' by the ratio of reduced masses for motion along the reaction
coordinate, $\mu_H^{\neq}/\mu_D^{\neq}$, and since the $\mu^{\neq}$'s contain only masses they are
not pressure dependent.[39] We would conclude then, that from the
semi-classical analysis of isotope effects, no changes due to
pressure would be expected.

V.   EFFECTS OF PRESSURE ON THE REACTION COORDINATE

The presence or absence of an isotope effect may depend upon
the details of the reaction pathway. The case of a two-step reaction,
the second being hydrogen transfer, provides a case in point. A
familiar example of this is in electrophilic aromatic substitution.
The nitration of benzene is understood to occur by the following

sequence.[40]

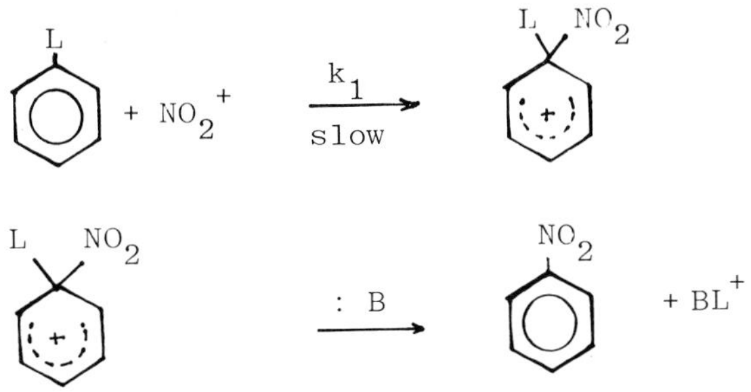

$$k_H/k_D = 1$$

L = H, D

The displacement of hydrogen and deuterium occur at the same rate
so the absence of the primary kinetic isotope effect is taken as
evidence that the first step is rate-determining. Diazocoupling,
sulphonation, and certain other reactions partake of the same
general characteristics of electrophilic substitutions (substituent
effects for example) but, in some cases at least, a kinetic isotope
effect is present.[41]

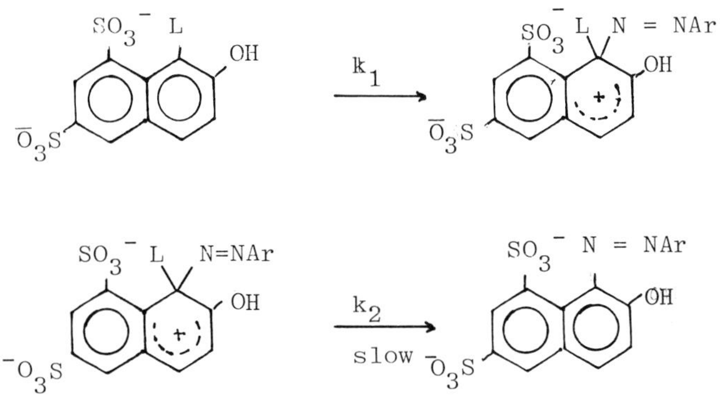

$$k_H/k_D = 3.5$$

The evidence is consistent with there being a slow or reversible
coordination of the diazonium ion to the aromatic ring followed by
a rate-determining loss of the proton. There is evidence that the
nitration of benzene is considerably accelerated by pressure (Table 1)

which presumably reflects the effect of pressure on the coordination
of the electrophile to benzene. This may be a general result
although few activation volumes for this reaction type have been
reported. The proton transfer, although between carbon and an oxygen
base, will have a much smaller volume of activation and may even be
positive as in the example cited above. One would conclude that
it would be possible, even likely, to shift the rate determining
process from the second to the first step of the reaction. Those
reactions which are of this mechanistic type and which show no
isotope effect at atmospheric pressure, would not be affected.
Those, however, which have an isotope effect for the reasons mentioned
should have it diminished under pressure as k_1 becomes comparable
with k_2. At present, no examples of this effect have been reported
although experiments are under way to ascertain the validity of
this analysis.

A second mechanistic feature which determines the magnitude
of an isotope effect is the extent of proton-transfer at the transition state. Many examples are now known for which the isotope
effect on the rate of an acid-base reaction reaches a maximum when
the two bases between which the proton is transferred are of equal
strength, $\Delta pK_a \sim 0$, and for which the standard free energy of transfer
also is zero. [42] It is inferred that the maximum isotope effect
is associated with a transition state in which the proton is shared
equally between the two bases. The same effect may be brought about
by changes of solvent; the hydroxide ion catalysed inversion of
(-)-menthone in aqueous dimethyl sulphoxide occurs with a maximum
isotope effect at a solvent composition of 30-40% dimethyl
sulphoxide. [43]

VI. QUANTUM MECHANICAL TUNNELLING

Non-classical behaviour of chemical systems may, in addition,
influence kinetic isotope effects. This is the phenomenon known
as 'tunnelling', a graphic illustration of the property whereby there
is a finite probability that a system may pass from the reagent state
to the product state though it possesses less than the classical
activation energy. Tunnelling is highly mass-dependent as it is
a manifestation of the Heisenberg Uncertainty Principle. [39] The
De Broglie wavelength, $\lambda = h/mv$, of very light particles can become
large compared with the dimensions of the activation barrier. In
such cases the probability of a molecule passing the activation

barrier is no longer two-valued as in the classical analysis (i.e. either 0 or 1 depending on whether $E < E_A$ or $E > E_A$) but becomes continuous, Fig. 6.   A significant contribution of tunnelling to a rate process is possible only for the movement of particles of very low mass. The electron, for example, frequently behaves non-classically and many solid-state electronic devices use this property. Hydrogen is the only atom for which non-classical behaviour need be considered and here, the tunnel correction (the ratio of the experimental rate to that calculated from classical considerations) will be largest for the proton. The transfer of hydrogen then, either as H^+, $H\cdot$ or H^-, may be subject to quantum mechanical tunnelling and may be detectable by kinetic methods if

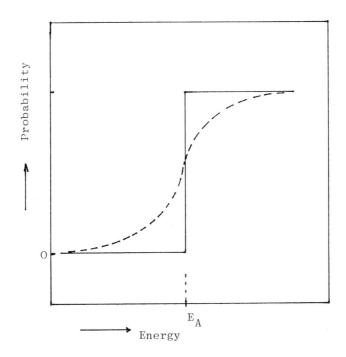

Fig.6. Probability of a particle passing a potential energy
       barrier as a function of its energy;   full line,
       classical function, dashed line, quantum mechanical function.

TABLE 4

Representative Reactions for which Tunnelling is Implicated.

Donor	Acceptor	K_H/K_D	A_D/A_H
H^+ transfers			
a. $pAnCL_2CH_2Br$	$t.BuO^-$ (E2)		10
b. $pAnCL_2CH_2\overset{+}{S}Me_2$	OH^- (E2)		8
c. $PhCL_2NO_2$	morpholine		10
d. $PhCL_2NO_2$	OH^-		2.4
e. $4-NO_2C_6H_4CL_2NO_2$	$HN=C(NMe_2)_2$	33	260
f. $4-NO_2C_6H_4CL_2NO_2$		21	9
g.	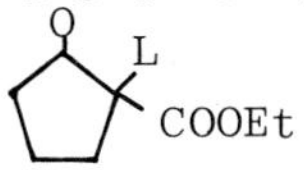 F^-	2.5	24
$H^\cdot$ transfers			
h. $CL_3CO_2^-$	$H\cdot$	22	
i. $PhCL_3$	$t-BuO^\cdot$	30	
j. $ArOL$	$-(CH_2\overset{\cdot}{C}HOAc)$	14-20	
k.			 10
H^- transfers			
l. $(4Me_2NC_6H_3)_3CL$	chloranil	11.5	24
m. $CF_3CL(OH)O^-$	MnO_4^-	14	2.4
n. $PLCLOH$ $\quad CF_3$	MnO_4^-	16	3.0

(L = H or D)

it occurs in a rate-determining step. The incidence of tunnelling
is, however, capricious. By no means all such reactions show non-
classical behaviour and, in fact, claims as to the importance of
this phenomenon have only been made for some dozen or so systems,
Table 4.[44]   The kinetic criteria by which tunnelling may be
recognised are threefold: abnormally large isotopic differences
in rates or activation energies or the failure of the Arrhenius
equation at low temperatures. To this may be added a fourth,
a pressure-dependent isotope effect, but it should be noted that
in any particular system not all these criteria are necessarily
satisfied simultaneously.

From classical considerations (equation 6), the maximum value
of the primary kinetic isotope effect, k_H/k_D, is about 8 at 25° [45]
so that experimental values greatly in excess of this are usually
cited as evidence for tunnelling. The reasoning behind this is that
the H reaction will be unusually fast if there is a tunnelling
contribution whereas the D analogue will be much less affected on
account of the greater mass of D compared with H. Values of
k_H/k_D as high as 20 at room temperature are known and much larger
values at low temperatures have been reported as will be shown below.

The extent of the tunnelling contribution is expressed by the
tunnel correction factor, Q:

$$Q = k_q/k_c \tag{7}$$

where k_q and k_c are rate constants based on quantum mechanical and
classical behaviour, respectively. It follows that k_q may be equated
with the experimental rate constant while k_c is the expected value
in the absence of tunnelling. Tunnel corrections can also be
estimated from the barrier (imaginary) vibrational frequency, $\nu^{\neq}$:

$$Q = \frac{\mu^{\neq}/2}{\sin \mu^{\neq}/2} \qquad \text{where} \quad \mu^{\neq} = \frac{h\nu^{\neq}}{kT} \tag{8}$$

and from the temperature-dependent behaviour of reaction rates.
This is usually expressed by the Arrhenius equation (eq. 9):

$$E_A = (\ln A - \ln k)RT \tag{9}$$

where R is the gas constant. The Arrhenius activation energy, E_A,
is a measure of the difference between the average energy of the
reacting molecules and the average energy of all molecules
present.   If tunnelling occurs the average energy of the
reacting molecules is lowered since those with less than the

classical activation energy will be capable of passing the barrier and the fraction of molecules potentially capable of reaction will not fall off with temperature as rapidly as classical theory would predict. Tunnelling therefore becomes of greater importance at lower temperatures and the Arrhenius plot of ln k against T^{-1}, normally linear over a wide range of temperatures, becomes curved at the low-temperature end, Fig. 7. The curvature is more pronounced for H-transfer than for D-transfer. Arrhenius plots of this type are taken to be evidence of rate-enhancement by tunnelling.

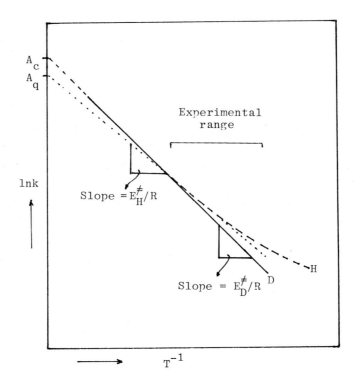

Fig. 7. Schematic Arrhenius plots for H-transfer (with tunnelling) compared with D-transfer (without tunnelling).

The intercept of the Arrhenius plot at $T^{-1} = 0$ on the ln k axis is the pre-exponential factor, ln A, a second parameter which can reveal the presence of tunnelling. From Fig.7 it will be seen that $\ln A_c > \ln A_q$. The relationship of these quantities with Q are as follows:

$$(E_c - E_q) = kT^2 \, d\ln Q \, / \, dt \tag{10}$$

$$\ln A_c - \ln A_q = \ln Q + T.d\ln Q/ \, dt \tag{11}$$

A-factors for isotopic hydrogen transfers in the absence of tunnelling differ slightly on account of vibrational differences, but the limits expressed by $0.5 < A_H/A_D < 2$ represent highly unusual systems on purely classical grounds. Therefore, values of A_H/A_D which lie much outside these limits, usually very high values, are interpreted as being due to tunnelling.

Solvation, the attractive force between solute and solvent, is ubiquitous and is certainly to be found between a labile hydrogen and a polar solvent. It seems plausible that the hydrogen being transferred tends to carry solvent with it during the activation stage of the reaction or, in other words, solvent motion is co-ordinated with motion of the hydrogen. This has the effect of increasing the effective mass of the hydrogen and thus rendering tunnelling negligible. While it is difficult to obtain concrete evidence for this postulate, it may be significant that in some examples at least the structure in the vicinity of the reaction centre is capable of excluding solvent by the presence of bulky groups. If these groups are absent, so are the criteria for tunnelling. Proton transfer between 2-nitropropane and some pyridines is a case in point:

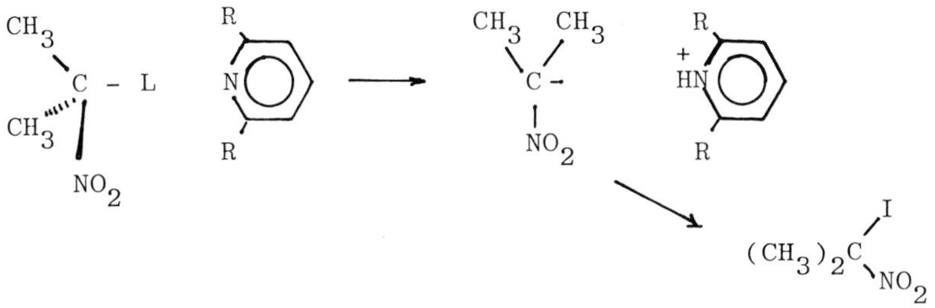

For R=H, $k_H/k_D = 7$ and no tunnelling is apparent. When R=Me, $k_H/k_D = 23$ which presumably indicates tunnelling at least for proton transfer. The _ortho_ methyl groups could plausibly prevent the close approach between solvent and acidic proton and may additionally affect the separation between carbon, hydrogen, and nitrogen in the transition state. Sterically hindered reactions are less exothermic and, by the Hammond postulate, partake of a later transition state than unhindered analogues.[47] We conclude that the incidence of tunnelling is related to some special feature of the transition state solvation or, rather, lack of it. From this it may seem that gas-phase reactions would be most likely candidates for tunnelling than those in solution. Many gas-phase hydrogen transfers, intramolecular or intermolecular, take place only at high temperature at which isotope effects are normally small and expected curvature of the Arrhenius plots slight. When suitable rates can be measured at low temperatures, tunnelling may be important. An example is the hydrogen atom exchange,

$$CF_3 \cdot + CHD_3 \begin{cases} \xrightarrow{k_H} CHF_3 + CD_3 \cdot \\ \xrightarrow{k_D} CDF_3 + CHD_2 \cdot \end{cases}$$

for which rates between 300 and 700K have been measured and a full vibrational analysis carried out.[46] Tunnel corrections are large at low temperatures, at which the isotope effect k_H/k_D can be as high as 18, and fall with increasing temperature.

Some recent theoretical work throws further uncertainty into the interpretation of anomalous Arrhenius parameters. Model studies indicate that high isotope effects and A-factors may arise in E2 eliminations (examples of which are known) as a result of an initial reversible proton transfer; that is, if the mechanism is more akin to the E1cB with appreciable return from the intermediate ion-pair to starting material.[73]

$$B + L + \overset{\backslash}{\underset{/}{C}}-CH_2Br \underset{k_{-1}}{\overset{k_1}{\rightleftharpoons}} \left[\overset{\backslash}{\underset{BL^+}{C^-}}-CH_2Br \right] \xrightarrow{k_2} \overset{\backslash}{\underset{/}{C}}=CH_2 + Br^- + BL^+$$

$$k_1 \sim k_{-1}$$

Other calculations using a force field model indicate that it is possible for the phenomena usually attributed to tunnelling to be generated when the hydrogen transfer occurs with a particularly 'loose' transition state,[74] a concept long recognised by Melander.[75] This would imply an unusually large separation between the two hydrogen acceptors in the transition state and could well be a situation brought about by steric crowding.

Since pressure is known to influence solvation, we consider
it likely that this in turn might alter the tunnel correction.
Any such effect might be expected to show up as a pressure-
dependence of the isotope effect or, equivalently, a difference in
volumes of activation for reactions of isotopic species since,
from equation 1 we obtain (eq. 12):

$$\frac{d(k_H/k_D)}{dP} = e^{(\Delta V_D^{\neq} - \Delta V_H^{\neq})/RT} \tag{12}$$

Several cases have now been investigated which test this postulate
and these are discussed below.

It is a striking fact that, of the large amount of kinetic
data relating to hydrogen transfer reactions which have been
published over the past fifty years, the number of systems for
which any evidence of tunnelling has been found remains extremely
small. Nor is there any clear common factor, such as structure,
reaction type, among those systems which show abnormal isotope
effects as may be seen from Table 4.   What reason can be given for
quantum mechanical behaviour not being the usual or at least a
common property of hydrogen transfers? The explanation, we believe,
lies in the nature of reactions in solution.

VII. EXAMPLES OF TUNNELLING IN HYDROGEN TRANSFER REACTIONS
(A) Hydrogen atom transfers to diphenylpicrylhydrazyl
     Palmer and Kelm[26] have measured rates of hydrogen transfer
from a series of phenols to the stable radical, diphenylpicrylhydrazyl:

Measurements have been extended to about 2 kbar whereupon rates increase moderately with volumes of activation in the range -11 to -13 $cm^3 mol^{-1}$. This must be the volume change associated with bringing together two neutral molecules close enough for H-transfer to occur and should not contain any electrostrictive term. Substituents in ortho-positions in the phenol markedly reduce rates, influencing mainly the A-factors as expected for a case of steric hindrance. The isotope effect for these reactions is remarkably variable, k_H/k_D ranging from 3 to 18 in closely related phenols, Table 5, and differing markedly in solvents benzene and toluene. No explanation of this variation has been proposed. However, of three cases in which values of both $\Delta V_H^{\neq}$ and $\Delta V_D^{\neq}$ are available, they are identical within experimental error. Consequently these isotope effects, even the abnormally high ones do not change with pressure.

TABLE 5

Volumes of Activation for Hydrogen Abstractions from phenols by DPPH.[26]

Phenol	$-\Delta V_H^{\neq}$ [a]	$-\Delta V_D^{\neq}$ [a]	k_H/k_D(1 bar)
2,4,6-trimethyl	13.5	–	–
2,4,6-triphenyl	12.3	–	–
2,4,6-tri(t-butyl)	13.7	13.6	2.4
2,6-di(t-butyl)	13.1	13.1	18.0
2,6-di(t-butyl)-4-methyl	13.1	13.2	3.0

a $cm^3 mol^{-1}$

(B) <u>Proton-transfer between Diphenyldiazomethane and Benzoic Acid</u>

This reaction has been extensively studied under a variety of conditions since it is clean and will take place in all solvents. At the same time it is conveniently followed by the disappearance of the red diazocompound.[48,49] The slow step is assumed to be the transfer of a proton from the acid to diphenyldiazomethane. The subsequently formed diazonium ion then loses nitrogen and the benzhydryl benzoate ion-pair collapses.

$Ph_2C=N_2$ $Ph_2C-N_2^+$ Ph_2CH $+$ N_2 Ph_2CH

$\underrightarrow{\text{slow}}$ $\longrightarrow$ $\longrightarrow$

The reaction is accompanied by a moderately large, though normal, isotope effect, $k_H/k_D \sim$ 4-5 and there is no indication that tunnelling is at all important. Rates in dibutyl ether under pressure have been measured[50], $\Delta V^{\neq} =$ -13 cm^3mol^{-1}, and the isotope effect is found, within experimental error, to undergo no change, Fig. 8.

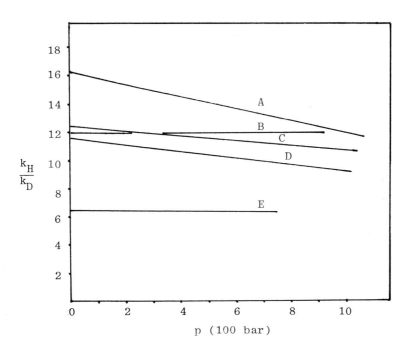

Fig. 8.   Effect of pressure on some Kinetic Isotope Effects: A, iodination of 2-nitropropane; B, hydrogen abstraction from tri-t-butylphenol; C, chloranil oxidation of leuco crystal violet; D, ionisation of p-nitrophenylnitromethane by tetramethylguanidine; E, reaction of diphenyldiazomethane with benzoic acid.

C. Proton Transfer Between p-Nitrophenylnitromethane and
   Tetramethylguanidine

Proton transfers from this nitro compound to a number of bases have been reported by Caldin and co-workers,[54-57] the product being an ion-pair.

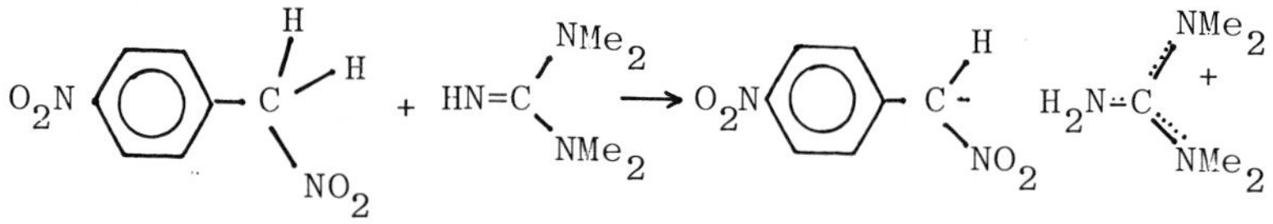

The reactions can be monitored by the change in visible absorption but, since they are rather fast, need to be followed by stopped-flow measurements. The kinetic characteristics of the reaction illustrated above are as follows: $k_H(25^{\circ}$ in anisole$)$ = 6070 $M^{-1}s^{-1}$; $\Delta H_H^{\neq}$ = -14.5 kcal mol^{-1}, $\Delta S_H^{\neq}$ = -39 cal $K^{-1}mol^{-1}$; $\Delta V_H^{\neq}$ = -17 cm^3mol^{-1}, k_H/k_D = 36. The enormous kinetic isotope effect testifies to the importance of tunnelling in this reaction. More recently, the reaction has been re-examined by Sugimoto, Sasaki and Osugi.[58] The volumes of activation in non-polar solvents are of the same order as those reported by Caldin, -14 to -18 cm^3mol^{-1} depending upon the solvent, although much smaller isotope effects were found. That the activation volumes are not more negative than these values is a consequence of the products being ion-pairs rather than free ions: ion association is known to be accompanied by a positive volume change, often very large,[6,21] and conversely in solvent dichloromethane volumes of activation for proton-transfer and for deuteron-transfer are essentially the same, -15 cm^3mol^{-1}, although the isotope effect is only 10.6. In toluene as solvent there is a difference in activation volumes for H-transfer and for D-transfer, $\Delta V_H^{\neq}$ = -19 and $\Delta V_D^{\neq}$ = -24 cm^3mol^{-1}, indicating an isotope effect which diminishes from 11.9 at 1 bar to 9.3 at 1000 bar. It is clear that the role of the solvent in all these reactions is crucial and the results, no doubt, are indicative of its complex nature. The reasons why the isotope effect is so solvent-dependent and why it is pressure-dependent only in certain solvents remain unexplained.

D. The Oxidation of Leuco-Crystal Violet

The colourless leuco-base of the dye crystal violet (I) is converted into the cationic purple dye, (II), by oxidising agents. A clean reaction occurs with chloranil (tetrachloro-p-benzoquinone, III) and it appears that the slow step is the transfer of a hydride ion.

    I   III   II

The reaction was investigated by Lewis and co-workers,[51,52] who showed that the isotope effect is abnormally large, k_H/k_D in acetonitrile at 25° being 12.3 and, besides, there is a very large discrepancy in the isotopic A-factors, A_H/A_D = 24. All these results clearly point to the importance of tunnelling. The volumes of activation for the oxidation of both isotopic species are very large and negative in accordance with the charge-separation which is brought about but they differ significantly, $\Delta V_H^{\neq}$ = -25 and $\Delta V_D^{\neq}$ = -35 cm^3mol^{-1}, which corresponds to a decrease of the isotope effect from 12 to about 9 at 2 kbar,[53] (Fig. 8). It is presumed that pressure brings about a change in the solvation of the transition state. Some measurements using isobutyronitrile as solvent were also made on the assumption that this would be a solvent of similar chemical properties, though greater bulk, than acetonitrile. It was found that the isotope effect diminished less rapidly with pressure supporting the notion that an increased solvent interaction with the hydrogen at elevated pressures is responsible for the reduction in isotope effect.

E. The Iodination of 2-Nitropropane

The substitution of the acidic hydrogen of 2-nitropropane by halogen is a base-catalysed reaction which has been extensively studied by Lewis and co-workers,[59,60] and by Bell and Goodall.[61] Under conditions such that the intermediate carbanion is immediately removed by iodine, the kinetic scheme can be represented as follows:

$$Me_2\overset{\displaystyle |}{\underset{\displaystyle NO_2}{C}}-H \; + \; :B \xrightarrow{\text{slow}} \; Me_2\overset{\displaystyle |}{\underset{\displaystyle NO_2}{C}}:^- \; \xrightarrow[I_2]{\text{fast}} \; Me_2\overset{\displaystyle |}{\underset{\displaystyle NO_2}{C}}-I \; + \; I^-$$

$$+ \; BH^+$$

The rate-determining step becomes the transfer of a proton from carbon to, usually, a nitrogen base. Large isotope effects are often observed: k_H/k_D with pyridine as base is about 10 but even larger values are obtained when pyridines containing alkyl groups in the 2,6-positions are used. An isotope effect of 24 has been reported and, although this may be an overestimate on account of the effect of atmospheric oxygen on the reaction[62], there seems no doubt that a value in the region of 16-20, depending somewhat on the solvent, is actually observed. That being so, a substantial tunnelling contribution is evidently observed. A large acceleration is brought about by pressure, $\Delta V^{\neq} = -31 \text{ cm}^3\text{mol}^{-1}$ in aqueous butanol, which is consistent with an ionogenic slow step, but in this case the volume of activation for iodination of the deuterated analogue is even more negative, $\Delta V^{\neq} = -40 \text{ cm}^3\text{mol}^{-1}$.[54] This corresponds to a decrease in the kinetic isotope effect from 16.5 at atmospheric pressure to about 11 at 1200 bar, Fig. 8.

The interpretation placed upon these results is that tunnelling is only important when a highly sterically hindered base is used. The ortho methyl groups may be considered to shield the proton under-going transfer from direct interaction with the solvent which otherwise would to some extent be coordinated to it and increase its effective mass. Pressure must decrease intramolecular distances and, apparently, bring about an increase in the proton solvation with a corresponding reduction in tunnelling.

F. The Oxidation of Benzhydrols by Permanganate

Rates of oxidation of secondary alcohols by alkaline permanganate increase rapidly with pH and show large primary kinetic isotope effects.[63]   It is concluded that the rate-determining step is the transfer of hydrogen from the alkoxide ion to permanganate, though whether this is a hydrogen atom or hydride transfer remains uncertain.   The final products are ketone and manganate.

$$Ph_2\overset{\displaystyle |}{\underset{\displaystyle |}{C}}\text{-H} \;+\; \overset{O}{\underset{O}{\diagdown Mn \diagup}}\overset{O^-}{\underset{O}{\diagup \diagdown}} \;\longrightarrow\; \begin{array}{l} Ph_2C\text{=}O^{\;\cdot} \;+\; HMnO_4^{\;-} \\ \qquad\qquad or \\ Ph_2C\text{=}O \;+\; HMnO_4^{\;=} \end{array}$$

The volume of activation for benzhydrol oxidation is -7 $cm^3 mol^{-1}$,[64] and the isotope effect is essentially normal, $k_H/k_D = 7$.   There is no reason to suspect any tunnelling contribution in this case and the volume of activation for D-transfer appears to be similar to that for H-transfer implying that pressure has no effect on the isotope effect.

The oxidation of trifluoroacetophenone would seem to be a likely candidate for an important tunnelling contribution since the reported isotope effect is abnormally high, $k_H/k_D = 16$ to 18 (depending somewhat on pH).[65]   This reaction is at present under investigation.

G. Intramolecular Hydrogen Transfer in Tri-t-butylphenyl

Highly hindered aryl radicals such as 2,4,6-tri-t-butylphenyl may be prepared in solution by the action of silyl radicals on the bromoarene or by photolysis of the corresponding peroxycarboxylate.

Stabilisation occurs spontaneously by internal transfer of one of
the eighteen equivalent hydrogens at quite a slow rate,
$(k_1 \sim 1 \text{ s}^{-1}$ at $-25^{\circ})$ so that the process can be followed by electron
spin resonance.[66,67] Rates may be measured at temperatures as low
as -250° and the same reaction may be followed using the totally
deuterated radical. This reaction provides one of the most
dramatic examples of tunnelling; isotope effects increase to the
largest known values as the temperature is lowered, approaching
10^4 at -250°. The Arrhenius plot is highly curved even for
deuterium transfer and H-transfer becomes almost temperature
independent below -100°, Fig. 9. This reaction must be considered

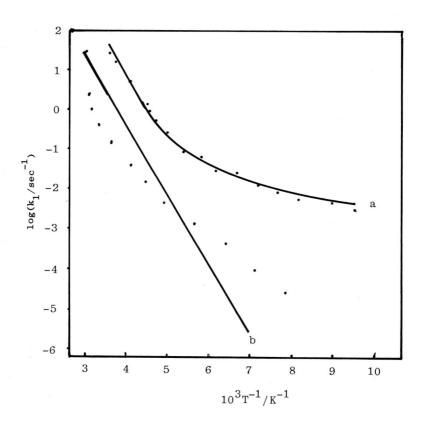

Fig. 9. Arrhenius plot for the isomerisation of 2,4,6-tri-t-
butylphenyl: (a) H-transfer;(b) D-transfer; solid lines
are calculated for a Gaussian activation barrier.[67]

a prime example of tunnelling in solution; similar conclusions
follow for reactions of analogous hindered aryl radicals. The
isotopic effects and their origin have been recently analysed by
LeRoy.[68]

Measurements have been made of the rates of isomerisation of
both H and D compounds under pressure in isopentane. Using a thick-
walled glass e.s.r. tube, pressures of several hundred bar could be
contained such as to enable volumes of activation to be measured:[69]

$$\Delta V_H^{\neq} = 5.3 \pm 1.7, \quad \Delta V_D^{\neq} = -1.2 \pm 2.0 \; cm^3 \; mol^{-1}.$$

Although experimental difficulties and the inability to reach
higher pressures left the uncertainties rather high, these values
appear to be significantly different, the deuterium transfer reaction
again having the more negative activation volume so that a reduction
of isotope effect under pressure is inferred.

H. The Bromination of 2-Carbethoxycyclopentanone

Enolisation of this β-ketoester is the initial and rate-
determining step in bromination, the reaction scheme being as follows:

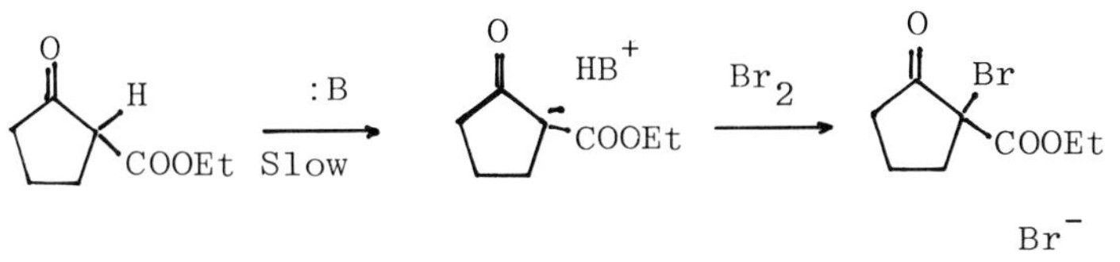

The enolate ion may be produced by a wide variety of bases and is
immediately scavenged by bromine. Reactions are usually carried out
with excess β-ketoester and with the base concentration buffered
so that the disappearance of bromine follows a zeroth-order rate
law. The reaction was intensively investigated by Bell and co-
workers who discovered several anomalies.[70] Certain bases reacted
with abnormally high values of A_H/A_D which the authors accepted as
good evidence for the involvement of tunnelling. For the water-
catalysed reaction A_H/A_D is only ∿2, just within the normal limits.

The fluoride-ion catalysed reaction, however, has A_H/A_D = 24 and E_D-E_H = 2.44 kcal mol^{-1}. Despite these figures, the kinetic isotope effects are all small, of the order k_H/k_D = 2-3, so that these values alone would give no indication of non-classical behaviour. Recently, values of the activation volumes of some of these reactions have been made.[71] Since they occur rather rapidly (of the order of 1 min. at 25° under zeroth-order conditions) stopped-flow techniques were necessary. The results show that, for the water-catalysed brominations (in 1:1 H_2O-D_2O mixtures) both $\Delta V_H^{\neq}$ and $\Delta V_D^{\neq}$ are -23 $\pm$ 1 cm^{3}mol^{-1} whereas a significant difference between these quantities exists for the fluoride ion-catalysed reaction, $\Delta V_H^{\neq}$ = -18 and $\Delta V_D^{\neq}$ = -27 cm^{3}mol^{-1}. Under these conditions, the isotope effect fell from 1.9 to 1.66 between 1 and 1000 bar. Further studies of catalysis by other bases which give intermediate values of A_H/A_D are in progress.

Conclusions

The effects of pressure upon primary isotope effects of eight reactions of widely differing type have now been examined and it seems clear that there is a link between the incidence of anomalous isotope effects or Arrhenius parameters on the one hand and isotopic differences in volumes of activation on the other. It is notable that in each case for which $\Delta V_H^{\neq} \neq \Delta V_D^{\neq}$, the latter appears to be the more negative quantity. If tunnelling, the usual explanation for the former anomalies is correct, then the pressure effect may constitute a further criterion of non-classical behaviour if one can be sure, at least, that the activation volume refers only to the hydrogen-transfer step throughout the pressure range. If this is so, then the most likely explanation of the reduction of isotope effects under pressure is by decreasing the importance of tunnelling and this in turn to the importance of the solvent. The Marcus approach to hydrogen transfer recognises a sequence of events depicted as follows:[73]

i	AH + B	$\overset{\rightarrow}{\leftarrow}$	AH ‖ B
ii	AH ‖ B	$\overset{\rightarrow}{\leftarrow}$	AH·B
iii	AH·B	$\overset{\rightarrow}{\leftarrow}$	A$^-$.HB$^+$
iv	A$^-$.HB$^+$	$\overset{\rightarrow}{\leftarrow}$	A$^-$ ‖ HB$^+$
v	A$^-$ ‖ HB$^+$	$\overset{\rightarrow}{\leftarrow}$	A$^-$ + HB$^+$

step (i) represents the formation of the encounter complex in which
hydrogen donor and acceptor approach each other, and (ii) represents
the formation of the reaction complex in which the two are oriented
for reaction and solvent reorganisation has begun. Step (iii) is the
actual hydrogen transfer and the only one in which tunnelling can be
involved. The reaction is completed by the symmetrical sequence of
events as the products separate. The volume of activation of the
reaction may be compounded of volume changes for several of these
steps, those which contribute depending upon the relative rates of
each in such a scheme. It is conceivable that pressure might
increase the importance of step (ii) at the expense of step (iii)
and thereby reduce the importance of tunnelling. Evidence is
available that the measured free energy of activation for proton
transfers contains a considerable proportion due to the free energy
of solvent reorganisation which we have referred to as electrostriction. The conversion of the encounter complex into the reaction
complex in particular would be a step in which solvent interactions
must be particularly important and one can devise electrostatic
models for coupling between motions of the proton and adjacent solvent
dipoles. Bell[39] has given for the effective mass of a proton,
m'_H moving, in the field of a dipole which has a moment of inertia I
at a distance ℓ:

$$m'_H = m_H + (I \sin^2 \theta / \ell^2) \tag{13}$$

where θ is the rotational angle induced in the dipole. As $m'_H > m_H$
the effective mass of the proton is increased by such coupling and
hence tunnelling is reduced.

On the other hand, such conclusions must be treated with some
reserve since alternative explanations to tunnelling have been
offered for abnormalities in both isotope effects and Arrhenius
factors. All hydrogen transfers are, in principle, subject to
internal return, that is, the reversibility of the transfer process
which is expressed by the equilibrium arrows in the scheme above.
A detailed kinetic analysis of such a complex scheme has been shown
capable of reproducing such anomalous behaviour.[74] Pressure, of course
would be expected to have a profound effect on the degree of internal
return (although the point does not seem to have been tested) in that
ionic products will be favoured over non-ionic reagents and hence
could reduce the degree of return. A further theoretical analysis
by McLennan using the Wolfsberg-Stern force field approach for

the calculation of isotope effects produces the conclusion that large isotope effects ($k_H/k_D > 17$) may result from simple proton transfers which go via very 'loose' symmetrical transition states.[75] This implies very low bond orders to the proton in the transition state and is evidently associated with higher than normal values of (E_D-E_H) and A_D/A_H.[76] It is less easy to understand how pressure could bring about a change in the molecular geometry of a transition state although a possible mechanism might be through an increase in the local dielectric constant of the solvent. This interpretation must therefore be borne in mind in a discussion of pressure effects.

It must be concluded that the reasons for abnormalities in $\Delta V^{\neq}$, A and k_H/k_D values are not without ambiguity and that perhaps different causes are operating in appropriate circumstances. It is likely that the kinetic effects of pressure can make a contribution to the understanding of such details in mechanistic chemistry.

REFERENCES

1. S.D.Hamann, 'High Pressure Chemistry and Physics', R.S.Bradley (Ed) (1963), Academic Press, London.

2. E.Whalley, 'The Use of Volumes of Activation for Determining Reaction Mechanism, Adv.Phys.Org.Chem. 2(1964)93.

3. W.J.Le Noble, 'Reactions in Solution under Pressure', Prog.Phys.Org.Chem., 5(1967)207.

4. K.E.Weale, 'Chemical Reactions at High Pressures', (1967), Spon. London.

5. G.Kohnstam, 'The Kinetic Effects of Pressure', Prog.Reaction Kinetics, 5(1970)335.

6. N.S.Isaacs, 'Liquid Phase High Pressure Chemistry', (1981), Wiley, Chichester.

7. F.T.Millero, Chem.Rev., 71(1971)147.

8. G.A.Lawrence and D.R.Stranks, Acc.Chem.Res., (1967)2173.

9. D.Palmer and H.Kelm (Eds.), 'High Pressure Chemistry'. (1978), Reidel, Amsterdam.

10. N.S.Isaacs and E. Rannala, J.Chem.Soc. (Perkin II), (1978)709.

11. B.T.Baliga and E.Whalley, J.Phys.Chem., 73(1969)654.

12. e.g. Nova Swiss, Effretikon, Switzerland; American Instrument Co., Silver Spring, Md.

13. H.Vanni, W.L.Earl and A.Merbach, J.Magnet.Res., 29(1978)11.

14. R.A.Grieger and C.A.Eckert, J.Amer.Chem.Soc., 92(1970)7149.

15. D.A.Palmer, H.Schmidt, R.Van Eldik and H.Kelm,
    Inorg.Chem.Acta, 29(1978)261.

16. K.Heremans, J.Snauwaert and J.Rijkenberg,
    Rev.Sci.Instr. 51(1980)806.

17. H.Lentz and S.O.Oh, High Temp.HighPres., 7(1975)91.

18. A.D.Yu, M.D.Wansbluth and R.A.Grieger,
    Rev.Sci.Instr., 44(1973)1390.

19. K.A.H.Heremans, ref. 9, p. 311.

20. W.J.Le Noble and F.Staub, J.Organometall.Chem., 156(1978)26.

21. G.Sweiton, J.von Jouanne and H.Kelm, Proc.4th AIRAPT Conf.,
    (Kyoto), (1975)652; F.K.Fleischmann and H.Kelm,
    Tet.Lett., (1973)3773.

23. L.M.Stock and H.C.Brown, Adv.Phys.Org.Chem., 1(1963)35.

24. M.Buback and H.Lendle, Z.Naturforsch, 34A(1979)1482.

25. Y.Ogo, M.Yokawa and T.Imoto, Makromol.Chem., 171(1973)123.

26. D.A.Palmer and H.Kelm, Austr.J.Chem., 30(1977)1229.

27. C.Brun and G.Jenner, Tetrahedron, 28(1972)3113.

28. J. von Jouanne, H.Kelm and R.Huisgen, J.Amer.Chem.Soc.,
    101(1979)151.

29. A.Sera, T.Miyazawa, T.Matsuda, Y.Togawa and K.Marayama,
    Bull.Chem.Soc. (Japan), 46(1973)3490.

30. S.D.Hamann, Austr.J.Chem., 28(1975)693.

31. H.Tiltscher and Y.K.Wang, Z.Physik.Chem. (Frankfurt),
    90(1976)299.

32. Y.Kondo, H.Tojima and N.Tokura, Bull.Chem.Soc.(Japan),
    40(1967)1408.

33. K.R.Brower, J.Amer.Chem.Soc., 85(1963)1401.

34. D.W.Coillet and S.D.Hamann, Trans.Farad.Soc., 57(1961)2231.

35. K.R.Brower and D.Hughes, J.Amer.Chem.Soc., 100(1978)7591.

36. D.M.Adams and S.T.Payne, Ann.Rep.Chem.Soc., A69(1972)3.

37. H.G.Drickamer and C.W.Franks, 'Electronic Transitions and
    the High Pressure Chemistry and Physics of Solids (1973),
    Chapman and Hall, London.

38. G.E.Wolrafer, J.Solution Chem., 2(1973)159.

39. R.P.Bell, 'The Tunnel Effect in Chemistry', (1980)
    Chapman and Hall, London.

40. K.Schofield, 'Aromatic Substitution', (1980), Cambridge U.P.

41. H.Zollinger, Adv.Phys.Org.Chem., 2(1963)163.

104

42. R.P.Bell, 'The Proton in Chemistry', 2nd Edition (1973), Chapman and Hall, London, p.265.

43. R.P.Bell and B.G.Cox, J.Chem.Soc.(B), (1970)194.

44. E.F.Caldin, Chem.Rev., 69(1969)135.

45. E.A.Halevi, Prog.Phys.Org.Chem., 1(1963)109.

46. T.E.Sharp and H.S.Johnston, J.Chem.Phys., 37(1962)1541.

47. G.Hammond,J.Amer.Chem.Soc., 77(1955)334.

48. N.B.Chapman, M.J.R.Dack, D.Newman, J.Shorter, and R.Wilkinson, J.Chem.Soc., Perkin II, 962(1974)971.

49. R.A.More O'Ferrall, Adv.Phys.Org.Chem., (1967)331.

50. N.S.Isaacs, K.Javaid, and E.Rannala, J.Chem.Soc.,Perkin II, (1978)709.

51. E.S.Lewis and J.K.Robinson, J.Amer.Chem.Soc., 90(1968)4337.

52. E.S.Lewis, J.M.Perry and R.H.Grinstein, J.Amer.Chem.Soc., 92(1970)899.

53. N.S.Isaacs, K.Javaid and E.Rannala, J.Chem.Soc., Perkin II, (1978)709.

54. E.F.Caldin and S.Mateo, J.Chem.Soc. (Chem.Comm), (1973)854; J.Chem.Soc., Faraday I, 71(1975)1876.

55. E.F.Caldin and G.Tomalin, Trans.Farad.Soc., 64(1968)2814.

56. E.F.CAldin, A.Jarczewski and K.T.Leffek, Trans.Farad.Soc., 67(1971)110.

57. E.F.C.ldin, D.M.Parboo, F.A.Walker, and C.J.Wilson, J.Chem.Soc.,Farad.I, 72(1976)1856.

58. N.Sugimoto, M.Sasaki and J.Osugi, Bull.Chem.Soc.Japan, (to be published).

59. E.S.Lewis and L.H.Funderburk, J.Amer.Chem.oc., 89(1967)2322.

60. E.S.Lewis and J.K.Robinson, J.Amer.Chem.Soc., 90(1968)4337.

61. R.P.Bell and D.Goodall, Proc.Roy.Soc.(A), 294(1966)273.

62. N.S.Isaacs and K.Javaid, J.Chem,Soc.Perkin II, (1979)1583.

63. K.Wiberg (Ed.), Oxidation in Organic Chemistry, (1965), Academic Press, New York.

64. N.S.Isaacs, unpublished work.

65. R.Stewart, J.Amer.Chem.oc., 79(1957)3057; R.Stewart and R.Van der Linden, Tetrahedron, (1961)211.

66. G.Brunton, D.Griller, L.R.C.Barclay and K.U.Ingold, J.Amer.Chem.Soc., 98(1976)6803.

67. G.Brunton, J.A.Gray, D.Griller, L.R.C.Barclay and K.U.Ingold, J.Amer.Chem.Soc., 100(1978)4197.

68. R.Le Roy, H.Murai and F.Williams, J.Am.Chem.Soc.,
102(1980)2325.

69. P.R.Marriott and D.Griller, J.Amer.Chem.Soc., in press.

70. R.P.Bell, R.D.Smith and L.A.Woodward, Proc.Roy.Soc.(A),
(1948)192.
R.P.Bell, J.A.Fendley and J.R.Hulett, Proc.Roy.Soc.(A),
235(1956)453.

71. N.S.Isaacs and K.A.H.Heremans, unpublished work.

72. H.Heydtmann, A.P.Schmidt and H.Hartmann,
Ber.Bunsenges, 70(1966)444.

73. R.A.Marcus, Techniques of Chemistry, Vol.VI, Part 1,
(1974), Wiley, New York.

74. H.F.Koch and D.B.Dahlberg, J.Amer.Chem.Soc., 102(1980)6102.

75. D.J.McLennan, Austra.J.Chem., 32(1979)1883.

76. L.Melander and N.A.Bergman, Acta.Chem.Scand., 30(1976)703.

Chapter 3

MAGNETIC ISOTOPE EFFECTS

NICHOLAS J. TURRO

Chemistry Department, Columbia University, New York, New York 10027

BERNHARD KRAEUTLER

Laboratorium fur Organische Chemie, Eidgenossische Technische Hochschule
Zurich, Switzerland

CONTENTS

INTRODUCTION

Today's organic chemist is usually quite familiar with the basic manifestations
and applications of isotope effects in his field of research. Studies involving the
use of isotopes have attained a high level of sophistication and reliability. Most
investigations have concentrated in two directions, namely, the use of isotopic substitution in organic chemistry to probe molecular properties in the ground state and
the transition state, and to elucidate reaction mechanisms in tracer work.[1-4] Follo
ing the widely applicable principle of the Born-Oppenheimer (BO) approximation[5-7]
where electronic and nuclear motion are treated separately, the same potential energ
surface is used for the description of molecules that differ only by isotopic substitution. As a consequence, the treatments of the whole apparent manifold of isoto
effects in chemical reactions (be they kinetic effects or equilibrium effects, prima
or secondary effects, steric, inductive or resonance effects, etc.) are usually
reduced to the problem of how different properties of the isotopic nuclei affect the
motions of a representative point on a single electronic energy surface. In general
the effect of the isotopic mass is sufficient to describe the perturbation caused
by the isotopic substitution in a chemical system.[6] Such unifying characteristics
of isotopic substitution are classified as "isotopic mass effects" (IME).

Under special circumstances, a second property (or nonproperty) of isotopic nucle
that is the presence (or absence) of a nuclear magnetic moment (nuclear spin), can

also lead to a second type of isotope effect on chemical reactions, classified as a "magnetic isotope effect" (MIE). This is a rather fascinating idea in view of the tiny interactions that exist between nuclear spins and possible magnetic fields (Table 1), but ample evidence now has been accumulated for the common occurrence of

TABLE 1

Correspondence between the value of a magnetic field and the associated splitting of the energy levels of an organic radical. Values of an order of magnitude only.

	Gauss	cm^{-1}	kcal/mole	sec^{-1}
very strong magnet	100,000 G	$10\ cm^{-1}$	10^{-2}kcal/mole	$3 \times 10^{11}\ sec^{-1}$
strong magnet	10,000 G	$1\ cm^{-1}$	10^{-3}kcal/mole	$3 \times 10^{10}\ sec^{-1}$
toy magnet, strong hf	100 G	$10^{-2}\ cm^{-1}$	10^{-5}kcal/mole	$3 \times 10^{8}\ sec^{-1}$
typical hf	10 G	$10^{-3}\ cm^{-1}$	10^{-6}kcal/mole	$3 \times 10^{7}\ sec^{-1}$
earth's magnetic field	1 G	$10^{-4}\ cm^{-1}$	10^{-7}kcal/mole	$3 \times 10^{6}\ sec^{-1}$
chemical bond	2×10^{8} G	$20,000\ cm^{-1}$	59 kcal/mole	$6 \times 10^{14}\ sec^{-1}$

MIE's under certain, special conditions.[8-10] Consequently, the classification into "isotopic mass effects" and "magnetic isotope effects" is appropriate, since two completely different physical mechanisms control each effect. The reader may be understandably reluctant to accept the possibility that magnetic properties might significantly influence the course of chemical reactions. Let us recall, however, the fact that magnetic resonance spectroscopy depends on the existence of nuclear magnetic moments. Nuclei with magnetic spin of 0 possess no magnetic moment and do not possess magnetic resonance spectra, i.e., ^{12}C with nuclear spin of 0 cannot be studied by NMR, but ^{13}C with a spin of 1/2 possess a finite magnetic moment and can be investigated by NMR. Table 2 lists the magnetic properties of the isotopes of three common elements contained in organic compounds (hydrogen, carbon, and oxygen).

Interestingly, the isotopic mass effect employs the motions of electrons and nuclei as its physical basis. On the other hand, the magnetic isotope effect employs the interaction between the nuclear and electronic spin motions which is absent in explaining the mass isotope effect.

In this chapter we present a brief, qualitative description of the theory of intersystem crossing (ISC) in radical pairs which, when considered in the framework of the molecular mechanics of a pair of organic radicals and the concept of cage reactions, provides a simple unified theoretical basis for understanding the magnetic (spin) isotope effect and magnetic field effects on chemical reactions. A set of criteria will then be developed, that will provide a protocol to help chemists

TABLE 2

Some magnetic properties of triads of common isotopes of some light elements.[a]

	γ[b]	μ_N[c]	I	Natural abundance (1%)
^{1}H	26,751	2.7927	1/2	99.985
^{2}H	4,106	0.8574	1	0.015
^{3}H	28,534	2.9788	1/2	---
^{12}C	0	0	0	98.89
^{13}C	6,726	0.7024	1/2	1.11
^{14}C	0	0	0	---
^{16}O	0	0	0	99.759
^{17}O	-3,627	-1.8937	5/2	0.037
^{18}O	0	0	0	0.204

(a) R.K. Harris and B.E. Mann, "NMR and the Periodic Table", Academic Press, New York, New York, 1978.

(b) Magnetogyric ratios ($S^{-1}G^{-1}$).

(c) Nuclear magnetic moment (erg/G).

---

distinguish between IME's and MIE's experimentally.

Examples will be given in which the magnetic isotope effect is used to separate ^{13}C (a magnetic isotope) from ^{12}C (non-magnetic isotope) and also to separate the magnetic isotope ^{17}O selectively from the non-magnetic isotopes ^{16}O and ^{18}O. The former case is noteworthy, since it will be seen that the efficiency of separation of ^{13}C from ^{12}C by use of a MIE mechanism surpasses considerably the separation efficiencies possible via a mechanism involving a conventional isotopic mass effect. Finally, the implication of MIE's will be discussed in terms of mechanistic studies, as well as with respect to the possibility of exploiting the MIE for the selective enrichment of heavy isotopes.

II. THE ISOTOPIC MASS EFFECT ON CHEMICAL REACTIONS - BRIEF RECAPITULATION.

The treatment of the effect of isotopic mass on chemical reaction kinetics and equilibria has been pioneered mainly by work laid down in the late 1940's.[11-12] The current status has been covered in excellent recent literature summaries.[1-4, 13]

Based on the BO approximation, the conventional interpretation of isotope effects on molecular properties reduced to the study of how different masses of isotopic nuclei affect the motion of a representative point on the same electronic potential energy surface. In turn, these studies are used as a tool to derive information about the potential energy surfaces on which the nuclei move, usually with special interest to probe the transition state structures.

According to the usual treatments, IME's can be traced back to a quantum mechanical effect; that is, to the quantization of the vibrational energy levels in a molecule. Qualitatively, based on the standard treatments within the "transition state theory,"[14] kinetic IME's (and similarly equilibrium IME's) are largely determined by the effect of isotopic masses on the differences of the zero point energy content of molecules, and IME's can be correlated with changes of crucial zero point vibrational force constants when one goes from the reactants to the transition state or products.

For most reactions near or below room temperature, thermally equilibrated molecules can be considered to be in the lowest vibronic and electronic state. This basis of the "low temperature approximation" for the treatment of kinetic IME's is due to the changes of the zero point vibration frequencies of a molecule upon isotopic substitution. The magnitude of their differential changes upon going from the reactant to the transition state determines the size and sign of the kinetic IME, which can be expressed, as in eq. 1 , as the ratio of the rate constants k/k* for the elementary steps for reaction of two molecules of different isotopic composition:

$$k/k^* \cong \sigma \cdot \exp\left[\frac{N \cdot h}{2RT} \left\{ \sum_i (\nu_i - \nu_i^*) - \sum_j (\nu_j - \nu_j^*) \right\}\right] \tag{1}$$

The asterisk (*) refers to the heavier isotope, ν_i and ν_j are the vibrational frequencies of the relevant modes in the reactant and transition state, respectively, σ is a symmetry factor, and R (gas constant), N (Avogadro's number), h (Planck's constant) and T (absolute temperature) have the usual meaning.

The "low temperature approximation" is usually sufficiently accurate to describe the kinetic IME for a reaction, so that a straightforward correlation with the empirical rate law, as given by the Arrhenius equation (eq. 2) is possible if only a single reaction step is considered:

$$k = A \exp(-E_a/RT) \tag{2}$$

and correspondingly,

$$k/k^* = A/A^* \exp\left[\frac{1}{RT} \cdot (E_a^* - E_a)\right] \tag{3}$$

We can therefore identify the kinetic IME with a difference of the activation energy of the reaction involving the molecules substituted with two different isotopes:

$$k/k^* \cong \exp\left[\frac{1}{RT} \cdot (E_a^* - E_a)\right] \tag{4}$$

i.e.,

$$A \cong A^*. \tag{5}$$

In experiments with hydrogen isotopes under conditions where the low temperature approximation is valid, eq. 4 and 5 have been found to be followed fairly broadly, with a ratio of the Arrhenius preexponential factors near unity.[13] Furthermore, even under adverse theoretical assumptions no mechanism for a significant deviation could be constructed, i.e., (A/A* = 1.0 $\pm$ 0.4).[13,15] Experiments with

isotopic rate results not compatible with eq. 4, involving hydrogen isotopes in
experiments with sufficiently low reaction temperatures, have been taken as
evidence for proton tunneling[13] or for more complex reaction kinetics than a
single rate determining step.[16,17]

Using the magnitudes of the common vibrational frequencies involved, estimates
for maximal differences of activation energies and, correspondingly, of upper
values for kinetic IME's, have been calculated and tabulated (Table 3).[12] These
values are presented here for illustrative purposes. They have been calculated
with extreme assumptions and in general these maximum IME's are larger by about
50% than those estimated more recently under seemingly more reasonable restrictions

TABLE 3

Estimated maximum ratios k/k* for rate constants at 25°C.[12]

	primary isotope effects	secondary isotope effects (for atoms bound to carbon)
k_H/k_D	18	0.46 – 1.74
k_H/k_T	60	0.33 – 2.20
k_{12_C}/k_{13_C}	1.25	0.98 – 1.01
k_{12_C}/k_{14_C}	1.50	0.97 – 1.02
k_{14_N}/k_{15_N}	1.14	
k_{16_O}/k_{18_O}	1.19	

In summary, we recapitulate the most characteristic properties of kinetic IME's
as the following:

(1) Kinetic IME's can in general be correlated with effects on the activation energy
of a chemical reaction and consequently their magnitude decreases considerably
with increasing temperature.

(2) Large kinetic IME's are characteristic for small atoms only, especially
hydrogen isotopes.

(3) Kinetic IME's are larger by about an order of magnitude, when they are primary,
(i.e., the reaction involves breaking or forming a bond to the isotopic atom),
than when they are secondary.

(4) IME's on chemical equilibria can reach nearly as large values as the corresponding kinetic isotope effects (due to their common physical basis).

(5) Sign and magnitude of a kinetic IME depend upon the underlying mechanism of
the reaction; they are normal, i.e., k/k* > 1, when a bond to the isotopic
atom is broken (or is looser in the activated complex).

III. THE MAGNETIC ISOTOPE EFFECT: A PERSPECTIVE

In general, discussions of isotope effects on chemical reactions omit any serious possibility of a magnetic (spin) isotope effect (MIE), i.e., an isotope effect due to a nuclear magnetic moment or nuclear spin. In principle, such effects could arise from an interaction of nuclear spins with a magnetic field,[18] such as that provided by the unpaired electrons in paramagnetic molecules or by applied laboratory magnetic fields. An order of magnitude estimate concerning the size of the energies involved in such interactions seems to justify the usual neglect of the effect of nuclear moments or spins on the dynamics of chemical reactions (Table 1). Even under the most favorable circumstances, interaction energies of only ca. 10^{-4} to 10^{-5} kcal/mole are not surpassed.

It is quite obvious that such tiny interactions, as caused by nuclear spins, cannot produce a significant effect on a chemical equilibrium. Such effects are still measurable, however, as demonstrated by the routine possibility to obtain nuclear magnetic resonance spectra. Furthermore, extending this argument to the treatment of reaction rates with the help of the "transition state theory" (where the virtual equilibrium between the reactant and the activated complex is considered), a mechanism for a significant MIE is not obvious. Qualitatively, the same conclusion can be drawn from the Arrhenius equation (eq. 2), when a possible effect of the nuclear spin on the activation energy is taken into consideration. Indeed, eq. 4 predicts the magnitude[10] of a kinetic MIE of not larger than about $\exp(0.001) \cong 1.001$.

Apparently, no physical basis for a significant MIE can be found, if only activation energies are considered. However, a second look at the Arrhenius rate law leaves an alternative basis for kinetic MIE in the form of the pre-exponential term A. Thus, if a rate determining transition occurs between states of different magnetic properties, small perturbations that help to cause transitions between such states might result in large effects on the A-factor of the Arrhenius expression for the apparent rate constant k. Specifically, as will be shown below, substantial magnetic effects on the A-factor are possible if such a transition involves states of different magnetic properties (i.e., triplet and singlet electronic states). In certain cases the tiny interactions as produced by nuclear moments and spins can be very effective in changing magnetic properties.

The attempt to treat a potential MIE with the rate expression derived from transition state theory can be reconciled with the fact that magnetic isotope effects are "symmetry effects". The usually neglected transmission coefficient κ can be used to account for the effect of nuclear spins on the probability of transitions between singlet and triplet electronic states (and vice versa; see reference 19 for treating κ in relation to intersystem crossing reactions).

IV. A DIRECT OBSERVATION OF THE MODIFICATION OF REACTION RATES BY NUCLEAR SPINS: CHEMICALLY INDUCED DYNAMIC NUCLEAR POLARIZATION

The influence of nuclear spins on the rates of specific chemical reactions was discovered in the form of the "chemically induced dynamic nuclear polarization" (CIDNP),[20,21] which later found a physical interpretation in the "radical pair theory".[22,23] Strong emission and abnormally large absorption signals appeared in the nmr spectra when radical reactions were taking place in the nmr probe. Quali- tatively, these abnormal signals are associated with deviations from the Boltzmann distribution of the nuclear spin states and can be explained as follows: the spins of the nuclei attached to the radicals can promote or inhibit the probability for a radical pair to recombine during an encounter. The products of the radical reac- tion in turn preserve the nuclear spin information for a sufficient time to be measured and recorded.   Thus, the appearance of CIDNP phenomena are evidence for radical reactions. A qualitative discussion of the physical basis for this phenomenon, as applies similarly to the MIE, is the subject of the subsequent paragraphs in Section V.

CIDNP actually thus provided the first experimental evidence for a MIE (with the sensitivity of the nmr spectrometer these effects could be measured, although the overall kinetic effect of the MIE was generally negligible). Soon thereafter the possibility of observing MIE's (as well as magnetic field effects) was pre- dicted.[24]

V. INTERSYSTEM CROSSING IN RADICAL PAIRS

As stated earlier, kinetic MIE's are expected if, in the rate determining step of a reaction, a transition between states of different magnetic properties occurs, such as from singlet to triplet electronic states (or vice versa). How can such an intersystem crossing (ISC) become influenced by the interaction with nuclear spins?

The Wigner Spin Conservation Rule states that, during an elementary chemical step, electronic and nuclear magnetic moments (spins) conserve their orientation.[25] For most organic systems, this rule forbids ISC from singlets to triplets or from triplets to singlets (S-T ISC   T-S ISC, respectively) in an elementary chemical step. The Spin Conservation Rule breaks down when an "elementary step" involves states whose lifetimes are long enough to allow the relatively slow ISC mechanisms, such as spin-orbit (SO) coupling and hyperfine coupling (hfc), to operate. Reac- tions involving radical pairs represent a large class for which the Spin Conser- vation Rule fails, i.e., for which electron and nuclear spins are not conserved. This collapse of the rule for organic radical pairs leads to several striking and unusual phenomena that are of interest to chemists:

(1) the observation of CIDNP:[20-23] a manifestation of different reaction effi- ciencies for radical pairs with a different population of nuclear spin states;

(2) the magnetic (spin) isotope effect on chemical reactions:[8-10,24] the observation of different reaction efficiencies in radical pairs that differ by
possessing nuclei with different nuclear magnetic moments and different
nuclear spins;

(3) the magnetic field effect on chemical reactions:[18,24-25] the observation of
a dependence of the reaction efficiency of a radical pair on the strength of
an externally applied magnetic field, especially at relatively weak field
strengths (<1000 Gauss).

(A) Some fundamental postulates about cage reactions involving radical pairs

Several useful postulates help to set a foundation for a basic understanding
of the mechanism of magnetic isotope effects (or in general of magnetic effects)
on the chemistry of radical pairs, with special consideration of organic molecules
(in particular those containing only hydrogen and second row elements).

Postulate 1: When the bond connecting two groups (R_1-R_2) undergoes homolytic
cleavage, the primary radical pair $R_1^{\cdot} R_2^{\cdot}$ is produced with complete
conservation of the spin state originally possessed by R_1-R_2 (in
accord with the Wigner Spin Conservation Rule).

Postulate 2: Singlet radical pairs can undergo "cage" reactions, such as recom-
binations and disproportionations; triplet radical pairs cannot
undergo cage reactions directly, but can do so only after intersystem
crossing to singlet radical pairs.

Postulate 3: The chemical reactivity of a radical pair may be controlled by the
hyperfine interactions of the (orbitally) uncoupled electrons of the
radicals of the pair with nuclear spins.

Postulate 4: The chemical reactivity of a radical pair, if controlled by the
hyperfine interactions, is modified by the application of a laboratory magnetic field.

(B) The vector model of spin mechanics

As a concrete example, we shall consider a triplet radical pair $^3(R_1^{\cdot} R_2^{\cdot})$ that
is produced (Postulate 1) via homolytic cleavage of a triplet precursor molecule
$^3(R_1-R_2)$. The intersystem crossing in the radical pair $^3(R_1^{\cdot} R_2^{\cdot}) \rightarrow {}^1(R_1^{\cdot} R_2^{\cdot})$ may
be influenced by MIE's (Postulate 3) and/or magnetic field effects (Postulate 4).
If this is the case, then the overall kinetics of the radical reaction can be
influenced by magnetic effects, since only $^1(R_1^{\cdot} R_2^{\cdot})$ radical pairs form cage products,
but $^3(R_1^{\cdot} R_2^{\cdot})$ cannot do so (Postulate 2).

The triplet radical pair can exist in one of three possible distinct states that
may be classified in terms of the magnetic sublevels of the triplet state, in contrast
to the (single) singlet radical pair state (with zero magnetic and spin angular
moment). These states may be visualized by considering the vector model of electron
spin (Figure 1).

116

THE STATES OF A SPIN CORRELATED RADICAL PAIR

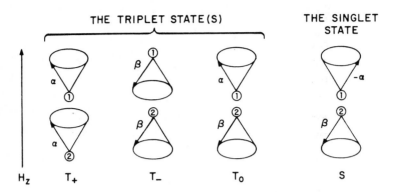

Fig. 1. Vector model of the triplet state and the singlet state of a radical pair (or a diradical). The direction of some arbitrary magnetic field H_z is employed as an orienting direction about which the electron spin vectors are imagined to precess. α refers to an "up" spin, and β refers to a "down" spin. The minus sign for α in the singlet state indicates that "up" spin is 180° out of phase with the associated β spin.

Let α symbolize the spin function for an electron spin that is pointing in the direction of an arbitrary magnetic field, H_z, and let β denote the spin function of an electron spin that is pointing in a direction opposed to H_z. The three triplet sublevels of a radical pair are labelled T_+, T_- and T_o, and have spin configurations $\alpha\alpha, \alpha\beta, \beta\beta$ and $\alpha\beta$, respectively. The singlet state, S, has spin configuration - $\alpha\beta$.

The T_o state has net spin angular momentum (as do T_+ and T_-) but the projection of this momentum along H_z is zero, i.e., the magnetic quantum number for T_o is $M_S=0$, while for T_+, $M_S=+1$ and for T_1, $M_S=-1$. The singlet state S of the radical pair is similar to the T_o sublevel of the triplet state in that it possesses an α spin and a β spin so that $M_S=0$. Thus, T_o and S both possess $M_S=0$. The (only) critical vectorial difference between T_o and S is due to the "phasing" of the spins. In T_o the vectors are 0° out of phase, whereas in S the vectors are 180° out of phase. The conversion of T_o to S (and vice versa) is a ISC which requires the "rewinding" of one spin vector relative to the other by 180°. Consequently, T_o to S transitions should be possible, when a mechanism exists that is capable of inducing a change of the phase relationship of the spin vectors.

Transitions between all three sublevels of the triplet state and the singlet

electronic state of a radical pair can be promoted by weak magnetic interactions
when the energy levels of the triplet states and the singlet state are quasi-dege-
nerate. This is possible, in general, when two conditions are fulfilled: (a)
the direct magnetic interaction between the electron spins is negligible (i.e.,
the electron exchange integral J is small), and (b) a strong external magnetic
field is absent. However, under the influence of a strong external magnetic
field, the triplet sublevels are split in energy away from S and only T_o remains
degenerate with S (both states possess a zero magnetic quantum number); i.e., in
a strong magnetic field, T_+ and T_- are raised and lowered respectively relative
to T_o by the amount of the Zeeman splitting, (Figure 2). The splitting of T_+ or

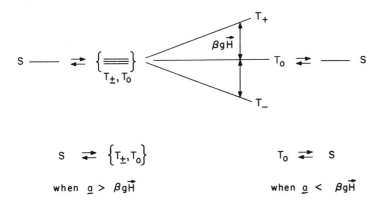

Fig. 2. Schematic representation of the Zeeman interaction $\beta g\vec{H}$ on the energetic
separation of T_+, T_- and T_o. When the Zeeman interaction is small
relative to other interactions (such as the hyperfine interaction whose
strength is given by $\underline{a}$, the hyperfine splitting constant), the triplet
and singlet states are energetically degenerate and all three triplet
sublevels interconvert with the singlet state. When $\beta g\vec{H}$ is large rela-
tive to $\underline{a}$, only $T_o \overset{\rightarrow}{\underset{\leftarrow}{}} S$ intersystem crossing occurs. The effect of $\beta g\vec{H}$
is to energetically split T_+ from S and thereby inhibit intersystem
crossing from or to these sublevels.

T_- from T_o (and S) is proportional to the magnitude of the applied field.[26]
    An important consequence of the application of a strong external magnetic
field is the inhibition of the $T_+ \to S$ and $T_- \to S$ transitions in the radical pair,
while $T_o \to S$ transitions are still probable (Figure 3). The former transitions
are inhibited because T_+ and T_- are no longer degenerate with each other or with
S, because of the Zeeman effect.
    In the absence of a magnetic field the spin vectors of the electron spins are
coupled to the strongest magnetic field present in the molecule R_1-R_2 (internal

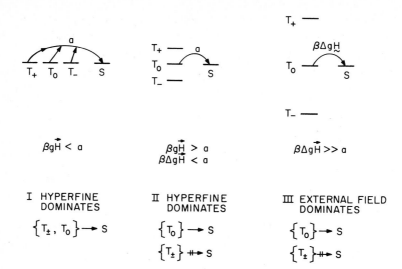

$$\beta g \vec{H} < a$$

$$\beta g \vec{H} > a$$
$$\beta \Delta g \vec{H} < a$$

$$\beta \Delta g \vec{H} >> a$$

I HYPERFINE DOMINATES

II HYPERFINE DOMINATES

III EXTERNAL FIELD DOMINATES

$$\{T_{\pm}, T_0\} \rightarrow S$$

$$\{T_0\} \rightarrow S$$

$$\{T_0\} \rightarrow S$$

$$\{T_{\pm}\} \nrightarrow S$$

$$\{T_{\pm}\} \nrightarrow S$$

Fig. 3. Representation of the three limiting cases for ISC in an organic radical pair. In cases I and II the hyperfine interaction (a) induces ISC. In case I, all three levels participate in ISC. In case II, only T_0 can undergo ISC. In case III, ISC is induced by the Zeeman interaction.

magnetic field), which they preserve during the bond breaking process. The spin vectors of the so-produced spin correlated radical pair ($\dot{R}_1 \dot{R}_2$) are then at some indefinite overall orientation, but at a well defined relative orientation or phase

In the absence of any magnetic perturbation on the electron spins of a spin correlated radical pair, the mutual orientation of spin momentum is maintained, and a singlet radical pair remains in the singlet electronic state (with net spin equal to zero), while a triplet pair keeps its net spin of one, i.e., remains a triplet.

(C) The mechanism of intersystem crossing in radical pairs

In the presence of several distinct magnetic fields, the correlated electronic spins of a triplet pair $\dot{R}_1 \dot{R}_2$ will tend to precess about the direction of the strongest field to which they can couple. If these fields are exactly the same for $\dot{R}_1$ and $\dot{R}_2$, then the precession rates ω_1 and ω_2 of the two spins are identical: in this case the difference in precessional rates $\Delta\omega$ is zero and the relative triplet phase is preserved forever. If, on the other hand, the magnetic fields experienced by the spins of $\dot{R}_1$ and $\dot{R}_2$ are different, then $\Delta\omega \neq 0$, the initial triplet phasing transforms into a "mixed" triplet-singlet phasing and eventually becomes a singlet phasing, i.e., a triplet-singlet intersystem crossing (ISC) occurs.

The qualitative dependence of the rate of $T_0 \rightarrow S$ conversions can be illustrated in terms of the vector model of magnetic moments obtained from classical physics.[27]

Let the Z axis correspond to a direction defining the orientation of the spin
vectors. The projections of the electron spins in the X-Y plane produce a vector
description of the time evolution of T_o and S that resembles a "spin clock" (Figure
4).[28] For a "pure" T_o state the spin clock only reads 12 o'clock, and for a "pure"

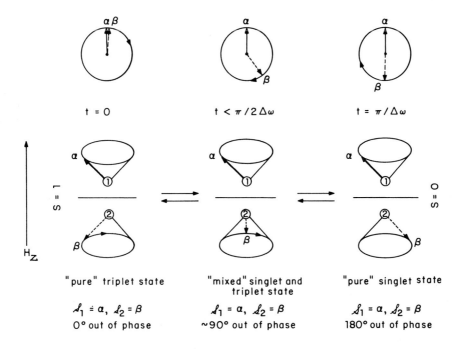

Fig. 4. A spin clock and vectorial representation of the $T_o \overset{\rightarrow}{\underset{\leftarrow}{}} S$ mechanism of
intersystem crossing in a radical pair.

S state the spin clock only reads 6 o'clock. When the electron exchange energy
(J) of the two unpaired electrons is large relative to the electron-nuclear hyperfine constant, a, i.e., when the radical centers are in close proximity (solvent
cage), the clock can read only 12 o'clock or 6 o'clock, i.e., T_o and S do not mix.
When J < a, i.e., when the radical centers are far apart (solvent separated),
T_o and S may mix. In such a situation the reading of the spin clock may correspond
to read any given time.

Suppose we start (t=0) with the spin clock reading 12 o'clock (e.g., the T_o
state of a triplet radical pair in a solvent cage) and then, as a result of
diffusive separation of the radical pair, we allow J to decrease to a value smaller
then a (e.g., the primary, geminate pair $\dot{R}_1$ and $\dot{R}_2$ diffuse out of the cage but
remain spin correlated). All that is now needed in order to convert T_o into S
(or S into T_o) is a rotation about the direction of H_z of one of the spin vectors
relative to the other. This will occur when the net magnetic torques on the spin

vectors are different. In addition to applied external fields, the spin vectors will experience magnetic fields arising from nearby nuclear spins and from electron orbital motion. Although the electron spins on $\dot{R}_1$ and $\dot{R}_2$ behave independently, the common origin provides a "spin correlation" or phase relationship between the spin vectors as they start to precess. The nearby nuclear spins make magnetic "communication" with the electron spins via electron-nuclear hyperfine coupling, whose magnitude is given by a, the hyperfine constant. The orbital motion of the electron makes magnetic communication with the electron spins via spin-orbit coupling, whose magnitude is reflected by the g-factor of the individual radicals.

(D) Time scales for intersystem crossing in radical pairs

Let us now consider the time scale for triplet-singlet rephasing in radical pairs. For concreteness consider first the conversion of T_o to S. The time τ_{TS} it takes T_o to convert to S equals the time required to rephase the two spin vectors by π radians. If $\Delta\omega$ equals the differential rate of precessions in radians/sec, then $\tau_{TS} = \pi/\Delta\omega$. We can also define a probability per unit time (analogous to a rate constant) for $T_o \rightarrow S$ interconversion as τ_{TS}^{-1}.

$$\tau_{TS} = \pi/\Delta\omega \text{ (Time for } T_o \rightarrow S \text{ rephasing)}$$

$$\text{(6)}$$

$$k_{TS} \equiv \tau_{TS}^{-1} = \Delta\omega/\pi \text{ (Rate for } T_o \rightarrow S \text{ rephasing)}$$

Suppose that the radicals $\dot{R}_1$ and $\dot{R}_2$ do not possess magnetic nuclei, but do possess different g-factors (g_1 for $\dot{R}_1$ and g_2 for $\dot{R}_2$). In this case, $\Delta\omega$ and k_{ST} are related to the laboratory magnetic field by the relationship,[18] where

$$\Delta\omega \overset{\sim}{=} |g_1 - g_2| \beta\underline{H} = \Delta g \beta\underline{H} \qquad (7)$$

$$k_{TS} \sim 3\times10^6 \ \Delta g\underline{H} \qquad (8)$$

$\underline{H}$ is in gauss and k_{TS} is in rad/sec.

For typical organic radicals $\Delta g \sim 0.001$. In the earth's magnetic field ($\underline{H} \sim 1G$), $k_{TS} \sim 3\times10^3 \ \text{sec}^{-1}$, a very small rate compared to the rates of other processes available to radical pairs. However, in very strong laboratory fields ($\underline{H} \sim 100,000 \ G$) $k_{TS} \sim 3\times10^8 \ \text{sec}^{-1}$. In this case, external field induced intersystem crossing may become competitive with other processes available to the triplet radical pair.

If $\dot{R}_1$ or $\dot{R}_2$ contains a magnetic nucleus whose hyperfine coupling constant is a, then k_{TS} will be proportional to some power of a even at zero external field. For an order of magnitude estimation[29]

$$k_{TS} \sim 3\times10^6 \ \underline{a} \qquad (9)$$

where $\underline{a}$ is in gauss and k_{TS} is in rad/sec. Typical values of $\underline{a}$ for organic radicals fall in the range 10-100 G, so that $k_{TS} \sim 3 \times 10^7 - 3 \times 10^8$ rad/sec for a single nucleus-electron hyperfine interaction. In this case, hyperfine induced intersystem crossing may become competitive with other processes available to the triplet radical pair.

(E) A general qualitative theory of CIDNP, magnetic isotope effects and magnetic field effects for reactions involving radical pairs.

The modern theory of CIDNP provides all of the essential concepts required for an understanding of magnetic isotope and magnetic field effects on reactions involving radical pairs.[30] Two important aspects of the theory involve a "sorting" principle and a "correlation" principle. These principles state that for a radical pair, nuclear spin states (i.e., +1/2 or -1/2 spins), electronic spin states (singlet or triplet) and triplet sublevels (T_+, T_- and T_o) may be "sorted" from one another if the pair can undergo geminate cage reactions of a "correlated" radical pair in competition with an escape process that either destroys the correlation or involves a reaction of the radical pair.

From eqs. 8 and 9 we learn that there are at least two mechanisms for ISC in a radical pair. The first (eq. 8) involves only a rephasing of the electron spin vectors and is brought about by an imbalance in the inherent precession rates of the vectors (Δg) and causes k_{ST} to increase as $\underline{H}$ increases. No nuclear spin flip is involved in this mechanism. Suppose, however, that $\dot{R}_1$ contains a magnetic nucleus of spin = +1/2 and a hyperfine coupling constant of $\underline{a}_1$, and $\dot{R}_2$ contains no magnetic nucleus. Eq. 8 then becomes

$$k_{ST} \sim 3 \times 10^6 \left| (\Delta g \underline{H} \pm 1/2\ \underline{a}_1) \right| \qquad (10)$$

Clearly, for a pair with $\Delta g > 0$ and if $\underline{a}_1$ is positive, then $k_{ST}^{+1/2} > k_{ST}^{-1/2}$ where the k's refer to radical pairs possessing nuclear spins +1/2 and -1/2, respectively. Since triplet radical pairs possessing nuclear spins of +1/2 undergo faster ISC than radical pairs possessing nuclear spins of -1/2, the former undergo cage reaction more efficiently than the latter if ISC is the rate limiting step for cage reaction. A mechanism for "sorting" nuclear spin states is available! As a result, the cage products are "enriched" in +1/2 nuclear spins and the escape products are enriched in -1/2 nuclear spins. We say that the nuclei of the cage and escape products are "polarized" (i.e., do not possess a Boltzmann distribution) and we have a rudimentary understanding of the theory of CIDNP.[22]

The second mechanism for ISC in a radical pair (eq. 9) requires a simultaneous flip of a nuclear spin and an electron spin. Clearly, this mechanism is impossible for a nucleus which does not possess a spin. We perceive that in the absence of a magnetic field ($\underline{H}$=0), radical pairs possessing magnetic nuclei can undergo ISC faster than radical pairs which possess only non-magnetic nuclei. A mechanism is thus available for sorting magnetic nuclei from non-magnetic nuclei! Table 2 lists the magnetic properties of some common light elements, hydrogen, carbon, and oxygen. Of the set. ^{12}C, ^{14}C, ^{16}O and ^{18}O possess a spin of zero and therefore are subject to magnetic isotope effects.

As mentioned earlier, application of an external magnetic field, in addition to increasing ISC via the Δg mechanism, energetically splits T_+ and T_- from T_o and S via Zeeman splitting of the triplet sublevels. T_o and S states remain degenerate if exchange integral J is negligible even when a strong laboratory field is applied (Figure 3). It is expected that when $\underline{H} > \underline{a}$, ISC from T_+ and T_- to S will become inefficient. As a result, when $\underline{H}$ is large enough, geminate cage reaction will occur predominantly from T_o via ISC to S. However, T_+ and T_- will still be able to undergo an "escape" process, which does not involve magnetic interactions. In effect, <u>application of a magnetic field provides a mechanism for "sorting" the chemistry of T_+ and T_- from that of T_o</u>!

(F) <u>Biradicals as "intramolecular radical pairs"</u>.

Homolytic cleavage of suitable cyclic organic compounds results in diradicals, that often can be treated similarly to a pair of (independent) radicals discussed above. Depending upon the length of the chain interconnecting the two radical centers (and the probability for an extended conformation) such diradicals may experience a small interaction between the unpaired electrons (the exchange interaction). In such a case, the dynamics of the evolution of the electronic spin state (the intersystem crossing) of the diradical can be closely related to the corresponding dynamics in a pair of structurally similar, but unconnected, radical. A residual average exchange coupling sometimes is taken into account to more close fit the experimental data (especially the magnetic field dependence of the signal intensities in CIDNP experiments).[30,31]

An important difference between a radical pair and the diradical is the inhibition of (diffusive) separation of the radical centers in the latter (beyond the separation of the most extended conformation); i.e., in diradicals the re-encounter of the chain ends is relatively probable, in contrast to the situation with a true radical pair. Indeed, the large probability for re-encounter should promote significantly the observation of MIE's in suitable radicals for which a small J is achievable.

VI. CRITERIA FOR THE OPERATION OF MAGNETIC ISOTOPE EFFECTS.

A kinetic isotope effect on a given reaction can be examined for the physical basis of its operation and characterized as an isotopic mass effect, or mainly a magnetic isotope effect, by considering the following criteria:[32]

(1) <u>The criterion of isotopic mass</u>:

Increasing the mass of an isotopic triad sequence of a single chemical element (e.g., ^1H, ^2H, ^3H, or ^{12}C, ^{13}C, ^{14}C, or ^{16}O, ^{17}O, ^{18}O, etc.) produces an easily predictable effect on an isotopic mass effect (roughly estimated via the reduced masses and as expressed in various diagnostic relationships).[12]

(2) The criterion of nuclear magnetic moment:

Using the same isotopic sequences (triad) of a single element as under (1), the effect of a magnetic isotope effect is easily predictable, but will scale as the size of the corresponding nuclear magnetic moment (which is e.g.,

$$\mu_{1_H} \sim \mu_{3_H} > \mu_{2_H} \text{ and } \mu_{13_C} > \mu_{12_C} = \mu_{14_C} = 0 \text{ and } \mu_{17_O} > \mu_{16_O} = \mu_{18_O}, \text{ etc}).$$

Thus, isotopic sequences can be chosen for analysis, as the ones above, that show completely different trends for isotopic mass and isotopic nuclear magnetic moment (see Table 2).

(3) The magnetic field criterion:

Quantitative analysis of the magnetic field effect on radical pair reactions predicts a strong dependence of the magnitude of a magnetic isotope effect on the applied external magnetic field. As a consequence of the competition between hyperfine induced and magnetic field induced intersystem crossing, a maximum of the kinetic isotope effect due to the nuclear spin is expected at relatively small external magnetic fields (in the range of the hfc-constants, i.e., at several hundred Gauss or smaller).[33] If an isotopic mass effect occurs in a radical pair reaction, such a maximum has no direct physical basis, although, of course, the modulation of the radical pair reactivity by the (strong) field could also cause an apparent, small (and smooth) perturbation on the isotopic mass effect.

(4) The hyperfine criterion:

Magnetic isotope effects correlate with the magnitude of the hyperfine coupling constants in the radicals (radical pairs) as they are available from esr measurements. No simple classification into "primary" or "secondary" magnetic isotope effects is, therefore, required. If different sites of a radical with non-zero spin density can be examined for isotopic enrichment, the effect goes parallel with the magnitude of the hyperfine coupling constants which can be significant at sites far from the point of initial homolysis, especially in delocalized radicals. In general, however, since in most cases the spin density in the radicals is highest on the atoms directly involved in the bond breaking/ making process, they usually have the largest hyperfine coupling constants and consequently are subject to the largest magnetic isotope effects.

(5) A few, less stringent criteria further exist:

    (a) the temperature effect is strong for an isotopic mass effect, which should greatly decrease with increasing temperature (since it can be correlated with a difference of activation energies of a reaction).

    (b) A direct single dependence of the magnetic isotope effect on the reaction temperature is not obvious, however.

    (c) The effect of reduced dimensionality, similar to the cage effect, in turn produced a much more pronounced influence on the magnetic isotope effect, than on the mass effect (for single step reactions, at least).

VII. THE MAGNETIC ISOTOPE EFFECT (MIE) AS A MECHANISTIC PROBE

(A) MIE's as a criterion for radicals as reaction intermediates

Magnetic effects on chemical reaction kinetics usually can be correlated with the occurrence of radical pairs (or biradicals) along the reaction pathway. To exclude possible magnetic effects on spin relaxation processes in heavier atoms,[34] such a correlation should be restricted to the discussion of organic chemical reactions involving the elements of the 1st and 2nd period.

Magnetic isotope effects, as a subclass of the magnetic effects, constitute evidence for the intermediacy of radicals (radical pairs or diradicals) as reaction intermediates in chemical processes. As a mechanistic criterion, the appearance of a MIE is equivalent to the observation of CIDNP. Studies concerning MIE's, if carefully performed to exclude possible isotopic mass effects, can, therefore, fully substitute for CIDNP investigations with respect to the mechanistic proof of the intermediacy of radicals. In some cases, where CIDNP is not observable for technical reasons, MIE's can be used as a complementary criterion.

Similar to Kaptein's rules[35] for CIDNP, for the observation of MIE's a simple "sign-rule" can be put forward (however, note that Kaptein's rules apply only to the situation at high fields):

(1) The depletion/enrichment of the magnetically (most) active nucleus can be correlated with the direction of the primary intersystem crossing of a geminate pair, and consequently with the electronic spin state of the nonradical precursor of the primary radical pair.

(2) A reactant, that acts as a singlet precursor, suffers depletion of its magnetically (most) active nucleus, and a reactant that acts as a triplet precursor shows complementary enrichment of its magnetically (most) active nucleus.

(3) The behavior of geminate cage products correlates with an efficiency that is opposite to that for escape products.

The above rules apply to experiments performed in the absence of strong external magnetic fields. Experiments under the influence of laboratory magnetic fields can be subjected to a complicated competition between the hyperfine and the Δg mechanisms, producing a less clear sign dependence.[10]

(B) The magnetic isotope effect as a dynamic probe.

MIE's can also be used to gain a more quantitative look at the dynamic processes of a radical pair. Significant MIE's require an efficient re-encounter of the spin correlated radical pair, after a diffusive separation has allowed hyperfine interactions to operate. MIE's, therefore, are sizable only when a significant cage effect (or a dynamic equivalent) is operative. Thus, the observation of MIE should be enhanced by restricting the reactions to reduced volumes of space and also in suitable intramolecular "radical pairs", as in some diradicals.

Magnetic isotope effects depend upon the competition of two reaction channels accessible to the radical pair, one of which contains the spin specific (cage) reactions and the other the (spin) unspecific (or escape) reactions. From a numerical evaluation of the time scale of the "spin clock", which allows one to estimate the rates of hyperfine induced intersystem crossing in the radical pair, the magnitude of the experimentally obtained MIE can be used to estimate the relevant rate of the escape. In homogeneous fluid solution, the rate of decay of the geminal spin correlated radical pair can often be correlated with the diffusive escape of the pair. If "irreversible" diffusive separation of a radical pair is inhibited by restricting the reaction space (on a molecular level), a chemical reaction can become the life limiting factor for the radical pair. MIE's then allow the estimation of the rates of such reactions (radical rearrangements and fragmentation reactions, etc.) in comparison with the relevant time domains schematically represented in Figure 5.

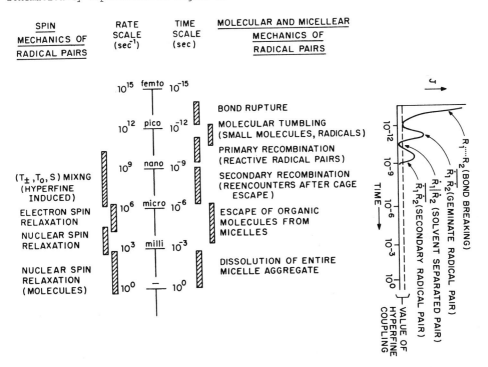

Fig. 5.   Time scales (order of magnitude) for the important dynamic processes
          involved in magnetic spin mechanics (left). These processes are compared
          to the time scales for important dynamic processes involved in molecular
          and micellar mechanics (right).

VIII. EXPERIMENTAL VERIFICATION OF THE OPERATION OF LARGE MAGNETIC ISOTOPE EFFECT

(A) Photolysis of dibenzyl ketone. Separation of ^{12}C from ^{13}C

   (1) Photolysis in homogeneous solution. Photolysis of $PhCH_2COCH_2Ph$ (DBK) in benzene solution at ambient temperature (λ_{ex} > 300 nm) results in a quantitative yield of CO and 1,2-diphenylethane.[36] The reaction pathway (Scheme 1) has been

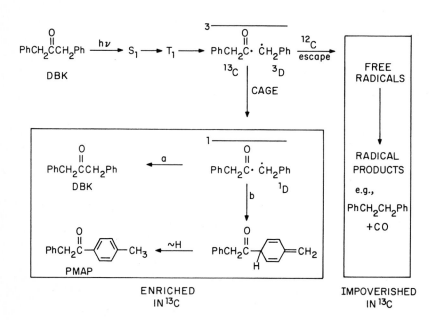

Scheme I.

demonstrated to involve homolytic α-cleavage of T_1 of DBK to produce a triplet radical pair ($PhCH_2\dot{C}O$ and $Ph\dot{C}H_2$) as primary products. The quantum yield of this reaction is high ($\phi \sim 0.7$). The small reaction inefficiency may be explained by assuming that a certain fraction of cage recombination to regenerate DBK is occurring. Indeed, CIDNP studies (^1H and ^{13}C) are consistent with a finite amount of cage recombination of $PhCH_2\dot{C}O$ and $Ph\dot{C}H_2$ radicals.[37]

   The pertinent ESR parameters[38] for the radical pair $PhCH_2\dot{C}O$ $\dot{C}H_2Ph$, produced from the photolysis of dibenzyl ketone, are listed in Figure 6. From the data, $\Delta g = 0.0018$ and the largest single value of $\underline{a}$ is 125 G for the $R\dot{C}=O$ carbon atom. If a dibenzyl ketone molecule contained ^{13}C at this carbon, we would expect k_{TS} of the triplet radical pair to be determined (at low $\underline{H}$) by the hyperfine interaction of ^{13}C (of $R\dot{C}O$) and the odd electron of the $PhCH_2\dot{C}O$. For a triplet radical pair containing ^{13}C corresponding to a natural abundance ($\sim$1.1%), k_{TS} will be determined by the ^1H hyperfine interactions of the $Ph\dot{C}H_2$ radical (note that the ^1H hyperfine interactions on the $PhCH_2\dot{C}O$ radical are $\sim$0 G). To the extent that

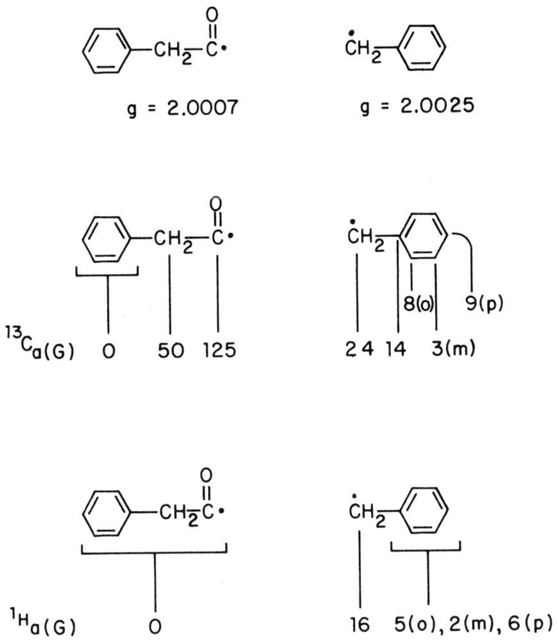

Fig. 6. ESR parameters of the $C_6H_5\dot{C}H_2$ and $C_6H_5CH_2\dot{C}O$ radicals. Ref. A. Berndt,
H. Fischer and H. Paul, "Magnetic Properties of Free Radicals," Londolt-
Bornstein, vol. 9, part b, Springer-Verlag, Berlin, 1977: $PhCH_2$ p. 543;
$PhCH_2\dot{C}O$, p. 321. The values for ^{13}C coupling of the benzene ring of
$C_6H_5\dot{C}H_2$ are estimated from spin densities of the benzyl radical and
measured proton hfc according to M. Karplus and G.K. Fraenkel, J. Chem.
Phys., 35, 1312 (1961) employing data from A. Carrington and I.C.P. Smith,
Molec. Phys., 9, 137 (1965).

radical pairs containing ^{13}C at the $R\dot{C}O$ carbon undergo more rapid $T_o \rightarrow S$ intersystem
crossing than those pairs possessing ^{12}C at the $R\dot{C}O$ carbon, the DBK regenerated by
cage reaction should be enriched in ^{13}C. The enrichment should be predominant at
the $R\dot{C}O$ carbon atom, since $\underline{a}$ (^{13}C) for $PhCH_2{}^{13}\dot{C}O$ is 125 G. However, a significant
enrichment of the CH_2 carbon of DBK should also occur, since the $\underline{a}$ (^{13}C) in
$Ph^{13}CH_2\dot{C}O$ is ∿50 G.

Experimentally, photolysis of DBK in benzene to very high conversions (∿99%)
followed by mass spectrometric (MS) analysis of the remaining DBK revealed that a
small enrichment in ^{13}C had occurred.[8,9] Employing DBK containing a natural
abundance of ^{13}C, the effect is barely outside of the experimental error. However,
if synthetically enriched DBK is employed, the measured enrichment which results
from photolysis is well outside of experimental error. The % enrichment is found

to depend on the extent of conversion. For example, starting with a sample initial containing 25.38% ^{13}C (total), at 45% conversion the % ^{13}C is 25.60%, and at 89% conversion it is 26.88%. This corresponds to a 5.6% net enrichment.[9] The magnetic isotope effect on chemical reactivity may be compared to the mass isotope effect on chemical reactivity by employing a single stage separation factor α, a quantity which is independent of the extent of conversion. For our purposes, α is a measure of the relative rate constants of the isotope species for an irreversible rate determining step.

$$\alpha = \frac{\text{rate of disappearance of } ^{12}\text{C compound}}{\text{rate of disappearance of } ^{13}\text{C compound}} \tag{11}$$

Bernstein[39] has derived a formula for determining α (for 1st order thermal reactions involving isotopes) from an appropriate plot of % enrichment as a function of % conversion (eq. 12):

$$\text{Log } S_f = [(\alpha-1)/\alpha] \cdot [\log\{(1-F_o^*)/(1-F_f^*)\} - \log(1-f)] \tag{12}$$

where F_o^* and F_f^* are, respectively, the molar fractions of ^{13}C in the starting ketone before reaction and after the fraction f of the starting ketone has been converted, and the overall enrichment factor S_f is given by

$$S_f = \frac{^{13}C_f/^{12}C_f}{^{13}C_o/^{12}C_o} \tag{13}$$

In equation 13, $^{13}C_f/^{12}C_f$ is the ratio of concentrations of ^{13}C and ^{12}C in the ketone after a fractional conversion f, and $^{13}C_o$ and $^{12}C_o$ are the corresponding concentrations in the starting ketone. For a small "single stage" enrichment factor α and relatively low ^{13}C contents, equation 12 can be approximated by equation 14.[39] Under the circumstances where equation 14 is applicable, it readily

$$\log S_f \overset{\sim}{=} -|(\alpha-1)/\alpha| \cdot \log(1-f) \tag{14}$$

allows one to deduce α from a plot of log S_f versus log (1-f). A typical plot is shown in Figure 7, for which the total ^{13}C enrichment in DBK after partial photolysis is monitored as a function of conversion f.

These results may be interpreted in terms of the mechanism given in Scheme 1. The enrichment parameter α may be correlated with an important photochemical parameter, the quantum yield, Φ, which could be verified experimentally (see below)[40,41], and allows the reformulation of equation 14 in terms of the disappearance quantum yields $^{12}\Phi$ and $^{13}\Phi$ for DBK containing no label and a ^{13}C label

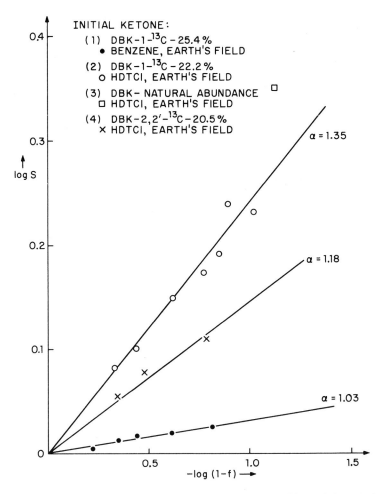

Fig. 7.   Typical experimental plots of the overall enrichment factor (S) versus
          the fraction of conversion (f) via eq. 14 for the photolysis of dibenzyl
          ketone in aqueous hexadecyltrimethylammonium chloride (HDTCl) solutions
          or in benzene solutions.

at a specific site (equation 15).

$$\log S_f \overset{\sim}{=} -\left|(^{12}\Phi - ^{13}\Phi)/^{12}\Phi\right| \cdot \log (1-f) \tag{15}$$

Applying Bernstein's formula to the photolysis of DBK in benzene yields, a
value of α= 1.03 ± 0.02.   The largest $^{12}C/^{13}C$ isotopic mass effects reported[12,57]
are of this magnitude.   Thus, although the ^{13}C enrichment is real, it cannot be
unambiguously associated with a magnetic isotope effect (based on the above observation only).

(2) <u>Experimental strategies to magnify the ^{13}C enrichment</u>. In analogy to the physical basis of CIDNP, the magnetic isotope effect originates in a reaction of radical pairs when for a key partitioning step an "escape" process competes with hyperfine induced intersystem crossing and eventual cage reactions. If escape is too fast and irreversible, then the number of molecules undergoing cage reaction is very small. In the case of DBK photolysis, the two most likely candidates for escape are diffusive separation and decarbonylation. $PhCH_2\overset{.}{C}O$ radicals can be trapped, and the lifetimes of these radicals has been estimated to be $\sim 5 \times 10^{-7}$ at room temperature.[42] On the other hand, escape of radical pairs from a solvent cage in benzene should take $\sim 10^{-10}$ s. Thus, in non-viscous homogeneous solution, diffusive separation from the initial solvent cage, a process that is required in order for hyperfine interactions to become effective, is followed by irreversible escape and formation of free radicals. If one can produce an environment for the triplet radical pair that (a) allows for diffusive separation to distances that allow J to "vanish", and for the hyperfine interaction to become effective, but (b) prevents or inhibits irreversible diffusive escape, then the competition between hyperfine induced intersystem crossing and escape will be modified and α should be increased.

(3) <u>The influence of viscosity on the ^{13}C enrichment</u>. From the above discussion it can be surmised that the observation of significant magnetic field effects on the reactions of radical pairs will require a proper detailed balance of time scales and the manipulation of various parameters such as diffusion coefficients of the radical pair, hyperfine interactions, $\Delta g\underline{H}$ interactions, and electron exchange interactions. Typical time scales and rates for these processes are summarized in Figures 5 and 8.

As the solvent viscosity is increased, the probability for recombination after a short diffusive escape of the primary radical pair should increase. To test this idea, photolysis of DBK at room temperature was conducted in solvents of a broad range of viscosity.[43,44] An increase of the enrichment factor α with increasing solvent viscosity was indeed observed. For example, upon proceeding from benzene ($\eta = 0.6$ cP) to dodecane ($\eta = 1.35$ cP), to cyclohexanol ($\eta = 30$ cP), and increase from 1.04 to 1.05 and to 1.07, respectively, for α was achieved.[43,44] Thus, although the enrichment increases as a function of viscosity, the effect is not of a very large magnitude.

(4) <u>The influence of micellization of ^{13}C enrichment</u>. For the operation of large magnetic isotope effects an ideal microenvironment will be of low viscosity (i.e., allows for rapid diffusive separation of a homolytically generated radical pair to sizable distances), but will possess a mechanism to induce efficient re-encounter of the radical pairs. In basic terms, we ask for the effect of an extended solvent cage, which contains the reacting species (molecules, radicals,

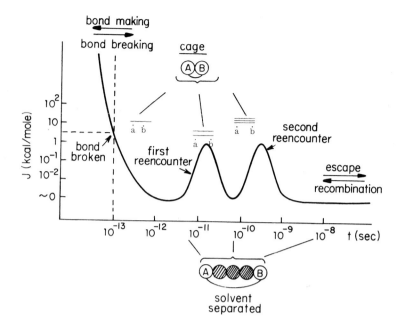

Fig. 8.   Variation of the magnitude of the exchange interaction J for a radical
pair as a function of time.

etc.) and possesses a reflecting boundary to send back the diffusing particles
towards a cage encounter (in a way similar to the effect of a long chain interconnecting the radical centers of a diradical).

A specific example of such a microenvironment of suitable dimension is that
provided by micellar aggregates.[45,46] These supermolecular structures are formed
spontaneously in sufficiently concentrated aqueous detergent solutions.[47] The
micellar "inside" represents a non-aqueous environment of hydrophobic character,[48,49]
corresponding to a dynamic reaction space of reduced dimensionality for a dissolved,
relatively unpolar, molecule or radical (pair). The polar micellar boundary surface acts as a reflecting boundary for these species. Such a static description
is adequate for rapid processes occurring on short time scales, where micellar
cages are effective (time domain < μsec - msec; i.e., the typical residence times
of solute molecules in micelles). Figure 9 provides a summary of structural
information for two commonly employed micelles, those consisting of hexadecyltrimethyl
ammonium bromide, HDTBr, (or chloride, HDTCl), and sodium dodecylsulfate, SDS.

The ideas employed in the DBK system are summarized schematically in Figure
10 in terms of potential energy surfaces.[50] The reaction coordinate represents
breaking of the $OC-CH_2$ bond. As the $OC-CH_2$ bond breaks the point "slides" down
the electronically repulsive triplet surface. When the bond is broken, the triplet surface becomes essentially degenerate with the ground state singlet surface.

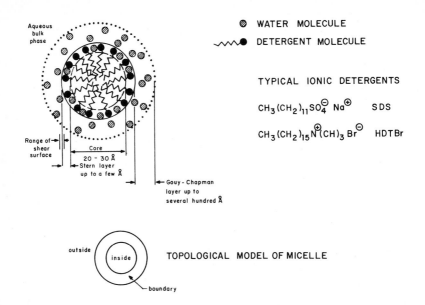

Fig. 9. Schematic model of the micelle aggregates produced in aqueous solutions of cationic detergents (e.g., HDTBr) or anionic detergents (e.g., SDS). It should be noted that the aggregates are dynamic species undergoing rapid monomer-micelle exchange (microsecond time domain).

A magnetic interaction is required before the representative point can make a "jump" from the T surface to the S surface. Such a jump can be induced by hyperfine interaction only when the point is far to the right, i.e., when the triplet and singlet are nearly degenerate and J < a. The "jump" from T to S requires a "hole" in the T surface, through which the point may fall.

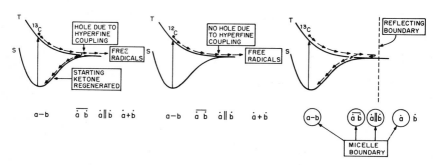

Fig. 10. Schematic potential energy surface representation of the photolysis of dibenzyl ketone. In the figure a = $C_6H_5CH_2CO$, and b = $C_6H_5CH_2$. See text for a discussion.

The role of the micelle may be viewed as providing a boundary which "reflects" the representative point back toward the hyperfine induced hole after an "overshoot" has occurred. Thus, diffusive escape is temporarily thwarted and a ^{13}C containing molecule receives extra chances to find a hole which allows return to ground state DBK. Eventually, of course, escape by decarbonylation will take place if neither diffusive escape nor bond formation occur.

In contrast to the results obtained from experiments in homogeneous solutions (of varying viscosity[43]), photolysis of DBK in aqueous micellar detergent solutions resulted in a dramatic increase of the ^{13}C isotopic enrichment.[9] For these experiments, it was useful to increase the level of significance by the use of ^{13}C labelled DBK (with artificially enriched ^{13}C content at the carbonyl or at the methylene carbons). For example, in a series of experiments, DBK-1-^{13}C-47.6% ^{13}C in the (carbonyl carbon) was used as the starting ketone. The recovered DBK from partial photolysis at room temperature of DBK-1-^{13}C-47.6% in aqueous, deaerated micellar solutions of the detergent hexadecyltrimethylammonium chloride (HDTCl), which was 0.05 M in HDTCl and 0.005 M in the ketone initially, was analyzed by several complementary methods (^{1}H-NMR, GC-MS, ^{13}C-NMR).

From mass spectrometric analysis of the recovered DBK (Figure 11), the fraction of DBK of m/e 211 (M+1 ion of DBK) increased progressively during the course of the photolysis of the detergent solution of DBK. After 91% conversion, the ^{13}C content in the carbonyl of DBK was determined to have increased from 47.6% to 63.7% (error limits $\pm$ 2%). The assignment of the progressive increase of the fraction of DBK with m/e 211 (from mass spectroscopic analysis) to the ^{13}C enrichment in the carbonyl of DBK is supported by complementary ^{1}H-NMR analysis of the recovered DBK at several stages of the experiment. A very sensitive test for the site of the ^{13}C enrichment (as the carbonyl of DBK) is available by the ^{13}C satellites in the ^{1}H-NMR spectrum: the methylene protons of DBK with a ^{12}C carbonyl

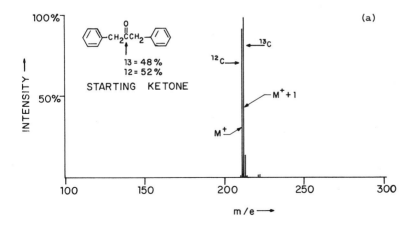

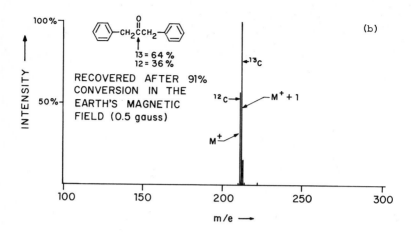

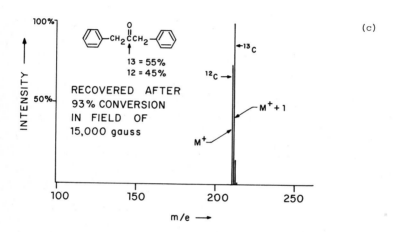

Fig. 11. Mass spectrometric analysis of dibenzyl ketone. (a) unphotolyzed ketone, synthetically enriched to the extent of 48% at the carbonyl carbon; (b) recovered ketone, after photolysis in the earth's magnetic field (0.5 G) to the extent of 91% conversion; (c) recovered ketone, after photolysis in a magnetic field of 15,000 G to the extent of 93% conversion.

are a singlet (at 3.66 ppm, $CDCl_3$, TMS as internal reference), while a doublet centered at the same chemical shift with $J_{13_{C,H}} = 6.5$ Hz is caused by ^{13}C-^1H coupling when the carbonyl carries a ^{13}C-carbon (Figure 12). Integration over the singlet and doublet signals allows the determination of the ^{13}C content of the carbonyl of DBK with good precision and accuracy to be $62 \pm 4\%$ ^{13}C (in the

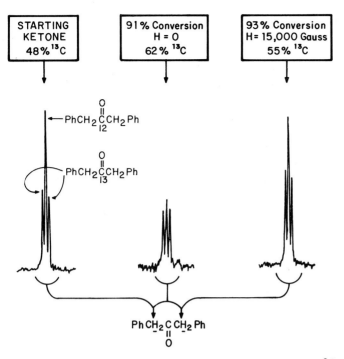

Fig. 12. ^{1}H-NMR spectra of DBK synthetically enriched in ^{13}C at C-1. Left: CH$_2$ absorption of starting ketone. Middle: CH$_2$ absorption of DBK recovered after 91% conversion by photolysis in HDTCl solution. Right: CH$_2$ absorption of DBK recovered after 93% conversion by photolysis in HDTCl solution which was subjected to a field of 15,000 G.

recovered ketone after 91% conversion), while the corresponding signal for the starting DBK corresponded to an original ^{13}C content of 48% in the carbonyl.

The DBK recovered after a 91% conversion, in the series of experiments with DBK-1-^{13}C-47.6%, showed an increased ^{13}C content of 63.7 $\pm$ 2% at its carbonyl, with a quantitative agreement between mass spectrometrically determined mass increase and the ^{13}C enrichment of the carbonyl of DBK, as analyzed via the "satellite method" of ^{1}H-NMR.

Using Bernstein's formula[39] (eq. 12), the single stage enrichment factor α_{CO}^{DBK} (for the carbonyl of DBK) can be calculated from the mass spectrometrically determined increase of the fraction of enriched DBK, to give $\alpha_{CO}^{DBK} = 1.35 \pm 0.03$ for room temperature experiments. Similarly, a linear plot of the overall enrichment factor S (see eq. 13) versus f, the fraction of conversion of DBK, gives a graphical representation of the approximate equation 14, with a slightly higher calculated enrichment factor $\alpha_{CO}^{DBK} = 1.41$. (This deviation of the single stage enrichment factor values is mainly a consequence of the high ^{13}C content of the starting material, see Figure 7).

To investigate the magnitude of the ^{13}C enrichment in DBK at sites other than the carbonyl, DBK-2,2'-^{13}C (^{13}C label at both methylene carbons of DBK) was used in enrichment studies, where photolysis of deaerated aqueous 0.05 M HDTCl solutions was performed as described above (at room temperature). Incomplete photolysis of DBK-2,2'-^{13}C-20.5% and recovery of DBK, followed by mass spectrometric analysis, established a substantial ^{13}C enrichment also in the CH_2 carbon atoms (under the conditions that a significant degree of label was present in the starting ketone). The corresponding single stage enrichment factor $\alpha_{CH_2}^{DBK}$ for the methylene carbons of DBK can be calculated from equation 12 to give $\alpha_{CH_2}^{DBK} = 1.18 \pm 0.03$ (for experiments at room temperature), see Figure 7.

The enrichment studies using ^{13}C labelled DBK as starting material, although more accurate, since characterized by a large absolute enrichment of the label during photolysis, also suffer from the fact that the small absolute enrichment of isotopes at other positions (other than the labelled atom) contribute to the measured enrichment. Therefore, photolysis was also performed with natural abundance DBK and the recovered ketone analyzed, as before, mass spectrometrically. Recovered DBK from partial photolysis in deaerated 0.05 M aqueous HDTCl solution showed a 200% increase of the ^{13}C content of a hypothetical single site of enrichment, after 95% conversion of the ketone, corresponding to a single stage enrichment factor of 1.68.[44] ^{13}C nmr analysis of this material revealed the carbonyl and methylene carbons to be the most efficiently enriched positions, but also indicated some enrichment at the aromatic carbon atoms. Indeed, the larger α values obtained by mass spectrometric analysis of the DBK recovered from photolysis of natural abundance DBK (and calculated for a single site of enrichment) should be taken as being due to a sum of (nearly) independent enrichments at all sites in the molecule. The summation of individual values gives a good correlation, i.e., $\alpha_{tot}^{DBK} = 1.68$, whereas $\alpha_{CO}^{DBK} = 1.35$, and $\alpha_{CH_2}^{DBK} = 1.18$. The enrichment pattern $\alpha_{CO}^{DBK} \gg \alpha_{CH_2}^{DBK} \gg \alpha_{other \ positions}^{DBK}$ agrees well with the ^{13}C nmr spectra of the recovered DBK from partial photolysis of natural abundance DBK in 0.05 M HDTCl solution. Furthermore, such a sequence of decreasing effect on the ^{13}C enrichment would also be expected from a comparison of the relevant hyperfine coupling constants in the radical pair generated by photofragmentation of DBK, the carbonyl carbon of the phenacetyl radical shows by far the largest coupling constant ($\underline{a}_{^{13}C}$ = 125 Gauss), followed by the methylene carbons in both radicals ($\underline{a}_{^{13}C}$ = 50 Gauss and $\underline{a}_{^{13}C}$ = 24 Gauss, respectively, in the phenacetyl and the benzyl radical). The aromatic carbon atoms show small coupling constants in the benzyl radical only (Figure 6). In this way the hyperfine criterion for the magnetic isotope effect is satisfied by the enrichment pattern.

(5) Cage effects and quantum yields for photolysis of dibenzyl ketones in

micellar solution. The "cage effect", that is, the percentage of geminate
radical pairs that undergo "cage" reaction, may vary from 0% (all geminate radicals
become scavengable free radicals) to 100% (all geminate radical pairs undergo "cage"
reactions). As mentioned earlier (Scheme 1) the photolysis of DBK in micellar
solution results in nearly quantitative formation (>90%) of 1,2-diphenylethane
(DPE) and a low yield (∿5%) of 4'-methylphenylacetophenone (PMPA). Addition of
$CuCl_2$ to the HDTCl solutions of DBK results in an initial lowering of the yield of
DPE followed by an independence of the yield of DPE upon further addition of $CuCl_2$.
New products, benzyl alcohol and benzylchloride are produced (Figure 13).

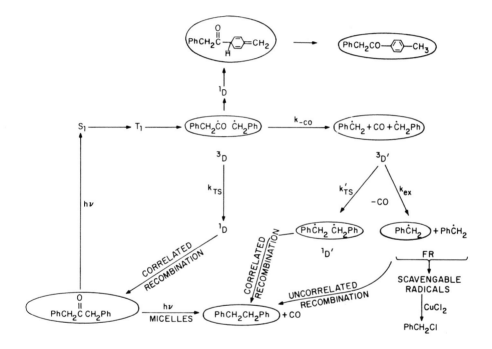

Fig. 13.   Reaction scheme for discussion of magnetic effects on the photochemistry
of DBK in micellar solution. The enriched species represent micellized
molecules or radicals.

Since HDTCl possesses a positively charged boundary, Cu^{2+} and $CuCl^+$, formed by
solution of $CuCl_2$ in the bulk aqueous phase, will experience a strong repulsion
as they approach the micellar boundary. As a result, $C_6H_5\dot{C}H_2$ radicals in micelles
are protected from reaction with Cu^{2+} and $CuCl^+$. However, upon escape from the
micellar to the bulk aqueous phase, $C_6H_5\dot{C}H_2$ radicals are quantitatively scavenged by
aqueous phase copper ions. From the effect of copper scavenging on the yield of

DPE, a cage effect of 30% is computed. The situation is schematically shown in Figure 13.

The cage effect for DBK in HDTCl may also be measured without employing a scavenger. When mixtures of $C_6H_5CH_2COCH_2C_6H_5$ and $C_6H_5CD_2COCD_2C_6H_5$ are photolyzed in HDTCl solutions (under conditions that micelles do not contain more than one DBK molecule), a mixture of isotopic DPE's is produced.[51] Experimentally, a cage effect of 30% is obtained by this method. Since the agreement is excellent between the measured cage effects by the scavenging method and by the product ratio method, we are confident that the use of copper as a scavenger does not significantly perturb the behavior of the radical reactions occurring in micelles of HDTCl.

Consider Figure 14 which shows schematically a situation in which a triplet radical pair is generated in a micelle. Suppose that ISC to form a singlet radical pair is determined only by hyperfine interaction, then at H=0 G, all three triplet levels will undergo hyperfine induced ISC to S, a singlet radical pair will be formed and cage reactions (e.g., combination) will occur. When a magnetic field is

IN THE PRESENCE OF SCAVENGER ONLY
GEMINATE PAIRS FORM R-R

WHEN H = O (THE EARTH's MAGNETIC FIELD)

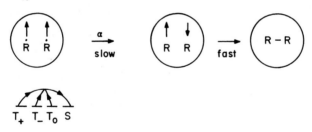

WHEN H > a

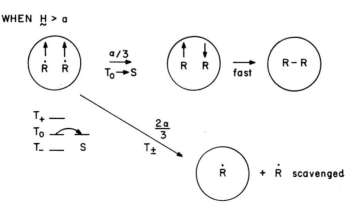

Fig. 14. Schematic representation of the influence of a laboratory magnetic

field on the efficiency of the cage reaction of a triplet radical pair in
a micelle. In the earth's field ISC from T_+ and T_0 to S is maximal and
the fraction α of a triplet radical pairs undergo cage combination. When
the applied field is strong enough to inhibit $T_+ \xrightleftharpoons{} S$ ICS, the fraction of
cage combination (in the limit) decreases to one third of the value at
zero field.

applied, T_+ will be split from S and when the field is strong enough, so that only
$T_0 \to S$ ISC will occur. These ideas predict the following experimental consequences:
(1) the cage effect will decrease, as a laboratory magnetic field is imposed on the
sample; (2) the effect of the field will "saturate" at a few hundred G, since hyperfine interactions are generally less than 100 G. Confirmation of these expectations
is shown in Figure 15.

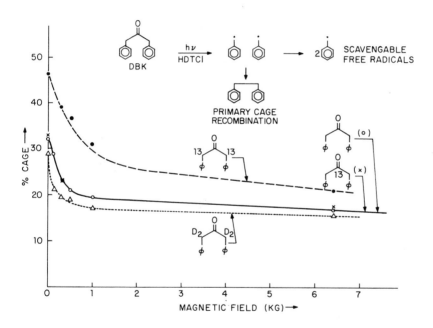

Fig. 15. Cage effects for DBK and isotopically substituted DBK's. Note that ^{13}C
substitution at C-1 has no influence on the cage effect, but ^{13}C substitution at C-2,2' causes a substantial increase in the cage effect. See the
text for a discussion.

Thus for the photolysis of DBK in HDTCl micelles, it can be seen that the cage
effect for DBK at $\underline{H}$=0 G, which is ∿30%, drops quickly to a value of ∿20% as fields
of a few hundred G are imposed on the sample, but as the field is increased from
∿500 G to ∿5000 G no further change in the cage effect occurs.[33] Figure 15

also demonstrates a remarkable magnetic isotope effect on the extent of cage reaction, i.e., for DBK which is enriched in ^{13}C in the 2,2' positions, the cage effect is dramatically higher (46%) than it is for natural abundance DBK (31%). This result is readily understood by postulating that triplet $C_6H_5{}^{13}\dot{C}H_2$ $^{13}\dot{C}H_2C_6H_5$ radical pairs undergo more rapid ISC than triplet $C_6H_5\dot{C}H_2$ $\dot{C}H_2C_6H_5$ radical pairs.

(6) Quantum yields for photolysis of dibenzyl ketone in micellar solution. Table 4 lists the quantum yields for photolysis of DBK and substituted DBK's. The

TABLE 4

Quantum yields for photolysis of dibenzyl ketones in 0.05 M HDTCl

Ketone (a)	Φ_{-K} (b)	$\dfrac{Cu(II)}{\Phi_{-K}}$ (c)	Cage effect
DBK	0.30	0.30	33%
DBK-1-^{13}C	0.22	0.22	33%
DBK-2,2'-^{13}C	0.25	0.25	46%
4-Me-DBK	0.23	0.23	52%
4,4'-di-Me-DBK	0.16	0.16	59%
4,4'-di-t-Bu-DBK	0.13	0.13	95%

(a) [Ketone] = 0.001 M except for [4,4'-di-t-Bu-DBK] = 0.0002 M.

(b) Quantum yield for disappearance of ketone. In the absence of Cu(II) the yield of diphenyl ethane(s) and the ketone isomer is nearly quantitative (>95%) for all ketones studied.

(c) Quantum yield for disappearance of ketone (-K) in the presence of 0.005 M CuCl$_2$.

quantum yield for disappearance of DBK (Φ_{-K}) in benzene, acetonitrile and other homogeneous organic solvents is ∿0.7. There is a substantial drop in Φ (to ∿0.3) in micellar solution. This decrease is probably associated at least in part with a more efficient recombination of the $C_6H_5CH_2\dot{C}O$ $\dot{C}H_2C_6H_5$ produced by α-cleavage.

The effects of isotopic and alkyl substitution on Φ_{-K} for ketone disappearance and for the amount of scavengable diaryl ethane are notable.[51] First, substitution of ^{13}C at the C-1 position leads to a decrease in Φ for net reaction, but does not lead to an increase in scavengable benzyl radicals. Consider Figure 15 for an examination of these effects. A magnetic isotope effect operates on the primary triplet $C_6H_5CH_2\dot{C}O$ $\dot{C}H_2C_6H_5$ pair and causes a more efficient recombination to occur for DBK-1^{13}C relative to DBK. This leads to a decrease in Φ_{-K} for ^{13}C at C-1. After decarbonylation the triplet benzyl radical pairs produced from DBK-1-^{13}C and DBK are equivalent, so that the cage effect is the same for each ketone.

On the other hand, substitution of ^{13}C at the α,α'-position leads to both a decrease in Φ_{-K} and to an increase in Φ_{+E}. From Figure 13, the decrease in Φ_{-K} is attributable to a magnetic isotope effect which leads to a more efficient recombination of $C_6H_5{}^{13}CH_2\dot{C}O$ $^{13}\dot{C}H_2C_6H_5$ radicals (due to more efficient ISC) relative to $C_6H_5CH_2\dot{C}O$ $\dot{C}H_2C_6H_5$ radicals. Likewise, the recombination of $C_6H_5{}^{13}\dot{C}H_2$ $^{13}\dot{C}H_2C_6H_5$ triplet radical pairs is more efficient than that of $C_6H_5\dot{C}H_2$ $\dot{C}H_2C_6H_5$ radical pairs, resulting in a larger value of Φ_{+E} for DBK-2,2'-^{13}C.

In the case of DBK's that possess alkyl substituents in the 4 (and 4') positions, the trend is for alkyl substituents to promote lower values of Φ_{-K} and larger values of the cage effect. The latter is readily understood in terms of a decreasing escape rate as the hydrophobicity of the benzyl radical is increased. The lowering of Φ_{-K} may be due to a combination of factors, including more efficient combination of geminate radical pairs before decarbonylation or a decrease in the efficiency of primary α-cleavage.

(7) Correlation of quantum yields with isotopic enrichment. A quantitative relationship has been established[52] between the value of α obtained from ^{13}C isotopic enrichment experiments and from quantum yield data for photolysis of DBK, as given in eq. 15. It is thus possible to measure α by two completely independent analytical methods: analysis of the ^{13}C content in the recovered DBK and determination of the enrichment, and conventional quantum yield measurements. The former do not involve knowledge of the light intensity, whereas the latter requires quantitative knowledge of the absolute light intensity. In Figure 16, the measured value of α as determined by these two methods is shown as a function of the applied magnetic field strength. The agreement between the two methods is excellent and serves as a convincing support for the mechanism given in Scheme I and eliminates the presence of artifacts related to optical properties of the system.

The absolute quantum yields for disappearance of DBK and for appearance of the photoproducts, 1,2-diphenylethane (DPE) and a novel product 4'-methylphenylacetophenone (PMAP), were measured at room temperature for benzene and aqueous HDTCl solution (0.05 M). PMAP was not observed as a product in benzene. In all cases the solutions were deoxygenated by nitrogen purging. The measured quantum yield for DBK (natural abundance, NA) disappearance ($\phi_{-DBK}^{12}{}^{C} = 0.72 \pm 0.12$) and the measured quantum yield for DBK-1-^{13}C-90% disappearance ($\phi_{-DBK}^{13}{}^{C} = 0.70 \pm 0.12$) are both large, but experimentally less than unity. Strikingly, the corresponding quantum yields for photolysis in aqueous 0.05 M HDTCl solution are much smaller: $\phi_{-DBK}^{12}{}^{C} = 0.30 \pm 0.03$ and $\phi_{-DBK}^{13}{}^{C} = 0.22 \pm 0.02$ (Table 3).

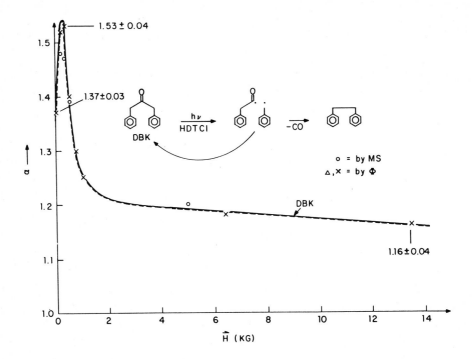

Fig. 16. Variation of the enrichment factor, α, as a function of magnetic field
strength.

(8) <u>Magnetic field effects on the</u> [superscript 13]<u>C isotopic enrichment during photolysis of</u>
<u>dibenzyl ketone</u>. As mentioned above, magnetic field dependencies of magnetic
isotope effects constitute one of their most important mechanistic criteria. For
the [superscript 13]C isotopic enrichment during the photolysis of DBK in micellar detergent solution, this enrichment was found to depend strongly upon the externally applied
magnetic field.

The photofragmentation of DBK produces a radical pair (benzyl and phenacyl
radicals) originally in the triplet electronic spin state. Recombination to ground
state DBK can occur only after intersystem crossing to a singlet radical pair state.
Application of an external magnetic field will decrease the rate of ISC of the
triplet radical pair, $^3(\dot{R}_1\ \dot{R}_2)$, because both T_+ and T_- will be split away energetically from S and will not be able to undergo efficient ISC (Figure 2). Thus, when
a strong external magnetic field is applied, chemical escape of the primary radical
pair, by decarbonylation, to give a (spin correlated) pair of benzyl radicals, will
compete more efficiently with a cage recombination to give back DBK (or to give
the isomeric PMAP).

The [superscript 13]C enrichment, therefore, could be considerably decreased in a strong external magnetic field (for two reasons):

(1) The two triplet sublevels T_+ and T_- are inhibited from ISC and, therefore, do not contribute to the recombination reaction that represents the mechanism for ^{13}C enrichment.

(2) The ISC from the remaining triplet level T_o is promoted not only by the hyperfine interaction between ^{13}C nuclear spins and the unpaired electrons (that brings about enrichment), but also the Δg-mechanism of ISC in radical pairs contribute to the rates of ISC (without involvement of nuclear spins), since the g-values of benzyl- and phenacetyl radicals differ significantly (Figure 6).

Experimentally, the effect of an externally applied magnetic field was tested by determining the dependence of the ^{13}C enrichment via the mass spectroscopically analyzed enrichment in recovered DBK as well as by measuring the absolute disappearance quantum yields for DBK and ^{13}C labelled DBK. As shown in Figure 16, the two independent sets of data agree quantitatively.

Strikingly, at the strongest magnetic field tested, (100,000 G not shown in Figure 16), the single state enrichment factor for the carbonyl carbon, $\alpha_{CO}^{DBK} = 1.02$, has decreased to a value of essentially no enrichment (or as typical of an isotopic mass effect for carbon isotopes). At smaller laboratory magnetic fields, the enrichment factor increases to $\alpha_{CO}^{DBK} = 1.16$ at a field of about 15 kGauss (Figures 11 and 12). Most importantly, the enrichment factor α_{CO}^{DBK} displays a maximum at a field of about 150 Gauss, where $\alpha_{CO}^{DBK} = 1.53$, that clearly identifies the origin of α as a magnetic isotope effect.

Such a maximum had been postulated to be significant for a magnetic isotope effect involving a single dominant hyperfine interaction (due to that nucleus) in a radical pair, based on quantitative simulations of the magnetic field dependence of hyperfine induced ISC in a radical pair.[26] Qualitatively its appearance and location on the magnetic field scale can be understood as being the result of a competition of the effect of a less efficient hyperfine induced ISC at increased magnetic field strength in general and the more rapid decrease of the fraction of radical pairs that undergo ISC due to the less efficient hfc with proton spins than with ^{13}C nuclear spins.

(9) ^{13}C enrichment in other "restricted spaces". The effect of the use of an aqueous micellar detergent solution of 0.05 M HDTCl was attributed to the existence of "micellar super cages" that restricted the reaction spaces available to the radical pairs formed by photofragmentation of DBK. This idea of the effect of "reduced dimensionality" on the radical pair reaction dynamics was tested (1) by correlating the single stage enrichment factor α_{CO}^{DBK} quantitatively with the relative concentration of micelles in aqueous solutions of the detergent HDTCl (Figure 17);[52] (2) by finding a substantial α_{CO}^{DBK} when photolysis of DBK is conducted in micellar detergent solutions of the related detergent sodium dodecyl sulfate (SDS): $\alpha_{CO}^{DBK} \simeq 1.4$;[44] (3) by arriving at qualitatively similar enrichment factors for

144

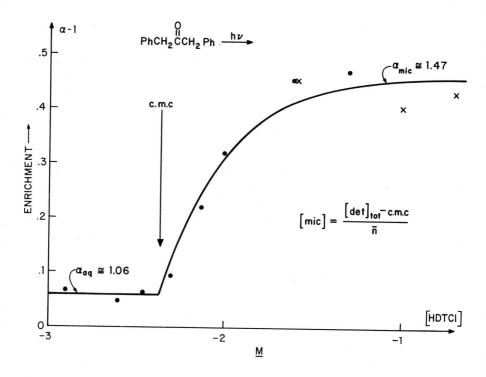

Fig. 17. Variation of the enrichment parameter, α, with concentration of HDTCl. ^{13}C enrichment of recovered DBK as a function of HDTCl. A sharp increase in α occurs at a concentration of HDTCl equal to the value $\sim 8 \times 10^{-3}$ M of the critical micelle concentration (CMC) of HDTCl in wate

^{13}C of the carbonyl of DBK, when photolysis of DBK is conducted in polymer films (e.g., polymethylmethacrylate) or in porous glass:[55] $\alpha_{CO}^{DBK} \sim 1.2$ to 1.3 were observed under these circumstances and again a strong externally applied magnetic field strongly decreased these single stage enrichment factors to $\alpha_{CO}^{DBK} \sim 1.1$.

In conclusion, it appears that the major effect of the micellar cavities on the dynamics of the photodecarbonylation of DBK and the concomitant enrichment of ^{13}C in the ketone are a consequence mainly (if not exclusively) of the constraint to a certain "restricted space", as is similarly also provided by the fluid portions of polymer films or the cavities of porous glass.

(10). Magnetic isotope effects as dynamic probes for radical reactions. The availability of a "spin clock" (Figure 4) due to the operation of magnetic isotope effects can be used as a tool for the mechanistic description of reaction mechanism involving radical pairs. To a good approximation the evolution of the spin motion of a radical pair usually can be semiquantitatively estimated from the hyperfine coupling constants of the radicals involved. Consequently, magnetic isotope effects can be used to act as probes for the time evolution of other processes, i.e., to investigate reaction mechanisms of radicals indirectly.

The discovery of a new photoproduct from the photodecarbonylation of DBK (in aqueous micellar detergent solution) proved to be of considerable value with respect to a more detailed description of the mechanism of this reaction. During the photolysis of DBK in aqueous micellar solution of HDTCl, the usual photoproduct, DPE, is accompanied by a new minor product, 4'-methylphenylacetophenone (PMAP), an isomer of DBK. PMAP is accessible formally from DBK by a 1,5-shift of the phenacetyl moiety on the remaining benzyl fragment, followed by tautomerization of the intermediacy formed cross conjugated hexatriene.

Interestingly, the formation of PMAP is accompanied by a significant increase of its ^{13}C content in the carbonyl, compared to the amount of ^{13}C in the carbonyl group of the starting material, DBK (Figure 18). In room temperature photolysis

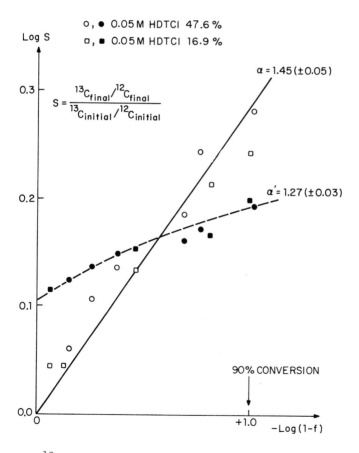

Fig. 18. ^{13}C content of recovered DBK and PMAP recovered from the photolysis of

DBK in aqueous 0.05 M HDTCl. After determination of α for DBK (experimental points, o, $\square$), the enrichment factor α' for PMAP was obtained from extrapolation of experimental points, ($\bullet$, $\blacksquare$) to zero conversion.

(27°C) of DBK in aqueous 0.05 M HDTCl solution, PMAP is formed with a 0.060 quantum yield from natural abundance DBK, but with a quantum yield of 0.074 from DBK-1-^{13}C.[52] The single stage enrichment factor α_{CO}^{PMAP} found from enrichment studies with DBK-1-^{13}C-47.6% (and as analyzed by mass spectrometry and ^{1}H-NMR satellite analysis, similar to the analysis for DBK itself), $\alpha_{CO}^{PMAP} = 1.22$, again correlates quantitatively with the quantum yield ratios, from which an enrichment factor $\alpha_{CO}^{PMAP} = 1.24$ was calculated.

Such a parallel isotopic enrichment in DBK and PMAP can be readily accounted for by considering the reaction mechanism of Scheme 1. The primary radical pair, a benzyl and phenacetyl radical pair, generated by photofragmentation of DBK effectively has two recombination channels available: recombination at the site of fragmentation (to give back DBK), and recombination at the para position of the benzyl radical, to end up, after tautomerization, in the isomeric PMAP. The apparent magnetic isotope effect, as manifested via the isotopic enrichment and/or the effect on the quantum yields, strongly support such a reaction mechanism with a common intermediate radical pair (generated in the triplet state) for both reaction channels. Studies on the magnetic field effect on the formation of PMAP as well as on the effect of the detergent concentration (i.e., the availability of micelles in the aqueous detergent solution) fully corroborate such a radical pair mechanism.[52]

More information concerning the kinetics of the photodecarbonylation of DBK in micellar detergent solutions was obtained by studying the effect of temperature on the ^{13}C enrichment and on the quantum yields:[52] The most important experimental findings are collected in Table 5, and are represented in Figure 19. Thus, increasing the temperature strongly increases the quantum yield for the disappearance of DBK (or of DBK-1-^{13}C). α_{CO}^{DBK} correspondingly decreases strongly (mass spectroscopic and quantum yield analysis). For PMAP, the efficiency of formation also decreases, but the concomitant enrichment of ^{13}C in the carbonyl becomes stronger at higher temperatures!

We can fully account for these experimental findings by considering the mechanism of Scheme 1, where the sole significant influence of temperature on the α's and the Φ's can be simulated by a temperature dependent rate of decarbonylation of the phenacetyl radical (using the numbers obtained[38] from esr-experiments in homogeneous solution studies).

From the experimental values obtained so far, the following relative efficiency and rate values for intersystem crossing of the phenacetyl-benzyl radical pair can be estimated:

(1) The intersystem crossing of a radical pair generated from DBK-1-^{13}C is more efficient by a factor of 1.24 (at room temperature) than the corresponding pair carrying a ^{12}C carbonyl. (This value can be obtained directly from the single stage enrichment factor for PMAP).

(2) As the most direct measure of the magnetic isotope effect of ^{13}C on the radical pair reaction in the photodecarbonylation of DBK, the rate of intersystem crossing for a pair from DBK-1-^{13}C is about two times faster than for the pair from DBK (with a ^{12}C carbonyl). This rate ratio is estimated to be nearly temperature independent in the range between 27°C and 70°C![52]

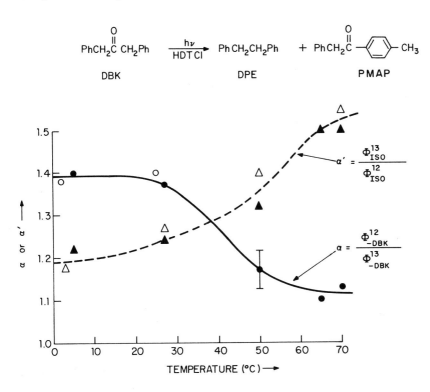

Fig. 19. Photolysis of ^{13}C-labelled dibenzylketone (0.005 M) in 0.05 M aqueous solution of hexadecyltrimethylammonium chloride (HDTCl) at various temperatures: single stage enrichment factors for starting material, α, and for the isomeric ketones, α', as determined using ^{13}C-enrichment studies (triangles) and using quantum yield values (circles).

TABLE 5

Temperature dependence of the quantum yields for photolysis of dibenzyl ketone.

Ketone	Temperature	ϕ_{-DBK}	ϕ_{PMAP}
DBK-NA	27°C	0.30 $\pm$ 0.01	0.060 $\pm$ 0.005
DBK-1-[13]C-90%	27°C	0.22 $\pm$ 0.01	0.074 $\pm$ 0.007
DBK-NA	70°C	0.75 $\pm$ 0.009	0.034 $\pm$ 0.003
DBK-1-[13]C-90%	70°C	0.67 $\pm$ 0.009	0.051 $\pm$ 0.002

(B) <u>Photochemistry of 1,2-diphenyl-2-methyl-1-propanone. Magnetic isotope and magnetic field effects</u>

The products of the photolysis of dibenzyl ketone are essentially the same for homogeneous solutions in organic solvents or for aqueous detergent solutions, since upon decarbonylation the benzyl radicals produced inevitably lead to DPE (in the absence of scavenger). In contrast, the photochemistry of 1,2-diphenyl-2-methyl-1-propanone (DPMP) depends significantly upon the solvent environment.[53] For example, photolysis of DPMP in CH_3CN leads to formation of styrene (ST) and dicumyl (DC) as the major products, with benzil and benzaldehyde formed as minor products. In micellar solution (e.g., HDTCl or SDS) the major products, in equal yields are styrene and benzaldehyde (Scheme 2). Dicumyl is not formed in detectable amounts.

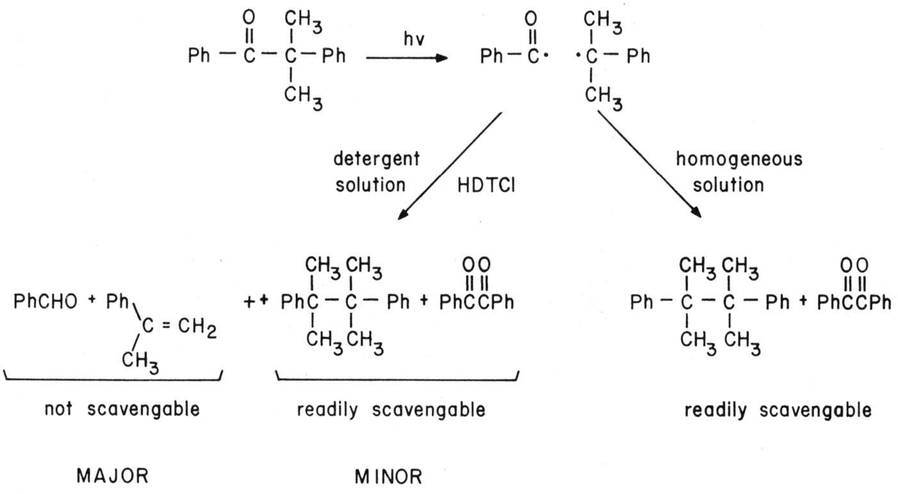

Scheme 2.

As in the case of dibenzyl ketone, the cage effect for photolysis of DPMP in micellar HDTCl solution may be computed from Cu(II) scavenging experiments. In fact, addition of Cu(II) does not affect the yield of the C_6H_5CHO or of $C_6H_5C(CH_3)=CH_2$ produced. Thus, the yield of either of these disproportionation products (since the compounds are produced in equal yield), based on the starting ketone consumed, equals the cage effect. The cage effects for DPMP and some of its isotopically substituted derivatives are given in Table 6.

TABLE 6

Cage effect on the photolysis of $C_6H_5COC(CH_3)_2C_6H_5$ in HDTCl solution.

Ketone	Cage Effect[a]	
	0 Gauss	1000 Gauss
$C_6H_5COC(CH_3)_2C_6H_5$	30	20
$C_6H_5COC(CD_3)_2C_6H_5$	23	13
$C_6H_5{}^{13}CO(CH_3)_2C_6H_5$	42	30

(a) The cage effect is defined as the % yield of C_6H_5CHO or $C_6H_5C(CH_3)=CH_2$ produced, based on starting material consumed.

Several aspects of the data in Table 6 warrant comment: (1) The cage effect is strikingly dependent on the isotopic composition of the ketone; (2) The cage effect decreases substantially when photolysis is conducted in a magnetic field of 1000 G (the major portion of the decrease occurs at fields strengths lower than 500 G); (3) Deuterium substitution (CH_3 groups) decreases the cage effect; (4) Carbon-13 substitution (CO carbon) increases the cage effect. These results are all qualitatively understandable in terms of the theory of magnetic isotope and magnetic field effects on correlated radical pairs.

First we postulate (Scheme 3) that in each case photolysis proceeds via T_1 which cleaves to produce a spin correlated triplet micelle caged radical pair $^3\overline{RP}$. This pair undergoes hyperfine induced ISC to eventuate in a singlet radical pair $^1\overline{RP}$ that undergoes disproportionation to C_6H_5CHO and $C_6H_5C(CH_3)=CH_2$. Escape of radicals from micelles competes only weakly for the parent ketone whose isotopic composition corresponds to natural abundance. In this case, ISC occurs mainly by the ^1H induced hfi of the methyl groups of the $C_6H_5\dot{C}(CH_3)_2$ radical (Figure 20). Deuterium substitution decreases the rate of ISC for $^3\overline{RP}$ because of the weaker hfi of the $C_6H_5\dot{C}(CD_3)_2$ radical. Substitution of ^{13}C from ^{12}C at the carbonyl carbon of the $C_6H_5\dot{C}O$ radical introduces a new and important hfi, i.e., the large ($\sim$125 G) hyperfine interaction of a single ^{13}C is as significant as the summation of proton hfi.

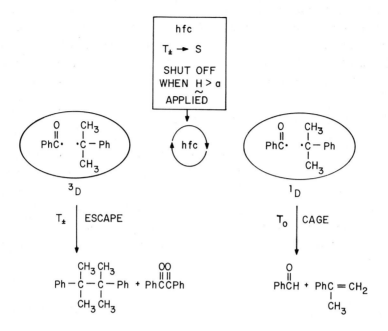

Scheme 3.

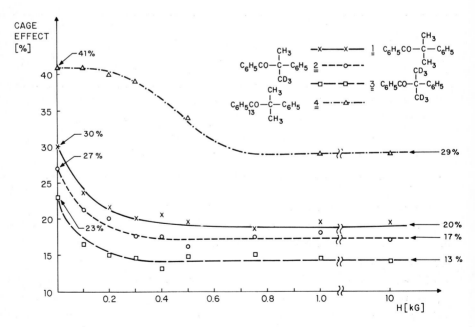

Fig. 20. The influence of isotopic substitution and magnetic field variation on the cage effect of 1,2-diphenyl-2-methyl-1-propanone.

The magnetic field effect in which the cage effect decreases with increasing field strength, is readily interpreted in terms of splitting of T_+ and T_- of $^3\overline{RP}$ from S and consequent decrease in the rate of $^3\overline{RP} \rightarrow {}^1\overline{RP}$. Since the rate of escape from the micelles is field independent, the net cage effect is reduced. The combination of isotope and magnetic field effects allow a variation of >300% in the degree of cage reaction!

(C) ^{13}C isotopic enrichment during the photolysis of other ketones in micellar detergent solution

Further support for the hyperfine mechanism for the ^{13}C enrichment in DBK is found in the observation that phenyl benzyl ketone (PBK) and phenyl adamantyl ketone (PAK) both are also strongly enriched in ^{13}C after partial photolysis of deaerated, aqueous (micellar) 0.05 M HDTCl solutions. Expressing the degree of enrichment in terms of Bernstein's single stage enrichment factor, α_{tot}^{ketone}, it is obvious that a similarly efficient effect is operating in these ketones as was found for DBK (Table 7).[44]

TABLE 7
^{13}C enrichment of ketones by photolysis in micellar solutions.[a]

Ketone	% Conversion	% Enrichment	α
DBK	95	200[b]	1.7[b]
PBK	90	140[b]	1.6[b]
PAK	95	220[c]	1.8[c]

(a) 0.005 M ketone, 0.05 M HDTCl, N_2 degassed, photolyzed through Pyrex with a Hanovia medium pressure lamp.
(b) Calculated assuming all ^{13}C enrichment occurred in one carbon.
(c) Calculated assuming ^{13}C enrichment distributed over two carbons.

The magnetic properties of the radical pairs formed by photofragmentation of PAK allowed one to expect a similar enrichment in each radical, i.e., in each moiety of PAK after recombination: 1-adamantyl radicals possess a large ^{13}C-hyperfine coupling constant at the 1-carbon atom (137 Gauss), as do benzoyl radicals at the carbonyl-carbon atom (128 Gauss). Indeed, degradation of the recovered PAK from partial photolysis in micellar solution revealed both moieties to be enriched to a very similar degree: 160% in the carbonyl carrying portion and 180% in the adamantyl fragment, compared to an average of 200% assuming selective enrichment of two carbon centers in the molecule.

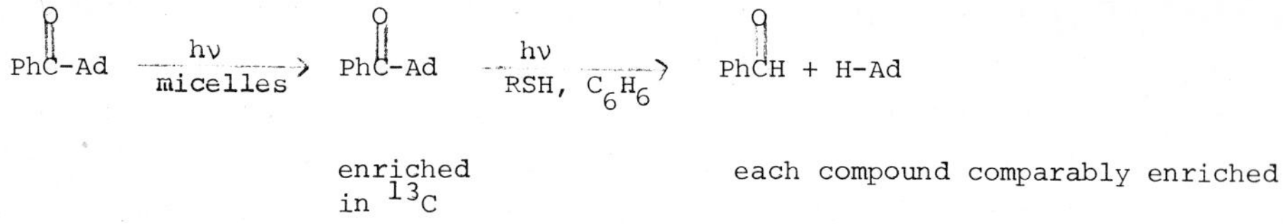

$$PhC(=O)-Ad \xrightarrow[\text{micelles}]{h\nu} PhC(=O)-Ad \xrightarrow[RSH, C_6H_6]{h\nu} PhCH + H-Ad$$

enriched
in ^{13}C each compound comparably enriched

(D) ^{13}C isotopic enrichment in the triplet sensitized photolysis of dibenzoyl peroxide.

Sagdeev and coworkers[10] have recently found a strong ^{13}C isotope effect in the triplet sensitized decomposition of dibenzoyl peroxide, which they correlate with the operation of a magnetic isotope effect. ^{13}C isotopic enrichment in the cage recombination product, phenyl benzoate, was established as follows (Scheme 4):

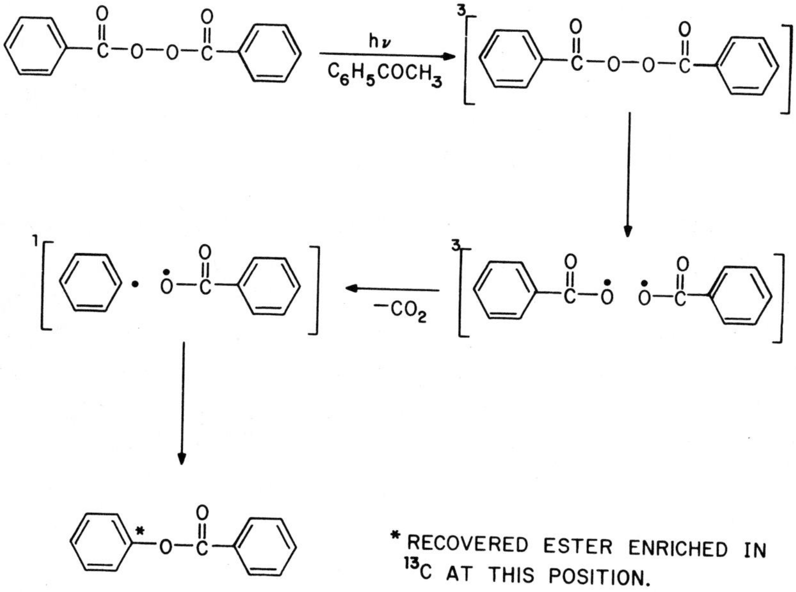

*RECOVERED ESTER ENRICHED IN ^{13}C AT THIS POSITION.

Scheme 4.

Irradiation of a CCl_4 solution of dibenzoyl peroxide and acetophenone to almost complete disappearance of the starting material resulted in a 2-3% yield of the phenyl benzoate. This was analyzed for isotopic carbon composition using ^{13}C nmr. From these measurements the amount of ^{13}C in the ipso position of the ester phenyl of phenyl benzoate was analyzed to have increased by 23 ± 5% (the abundance of ^{13}C in the other positions was found to be unchanged).

Theoretical estimates show that a roughly two-fold increase of the ^{13}C content in the 1-position of phenyl benzoate can be expected. As expected from theory, thermal decomposition of dibenzoyl peroxide resulted in no change of the content of ^{13}C in the 1-position of the cage product phenyl benzoate. The operation of a magnetic isotope effect thus seems highly probable. In the triplet sensitized decomposition, the primary radical pair is formed in the triplet electronic spin state. It has to intersystem cross to the singlet before recombination is possible with formation of the phenyl benzoate. The carbon at position 1 of the phenyl radical has a large hyperfine coupling constant, and therefore, should contribute most to a hyperfine induced ISC to the singlet state.

(E) Magnetic isotope and magnetic field effects on the formation of singlet oxygen from thermolysis of endoperoxides

The thermolysis of certain endoperoxides of aromatic compounds produces molecular oxygen quantitatively. From a study of activation parameters, it has been found that these reactions proceed via two pathways:[54,55] (1) a concerted mechanism in which $^1O_2$ is produced quantitatively, and (2) a diradical mechanism in which both $^3O_2$ and $^1O_2$ are produced. Magnetic field and magnetic isotope effects potentially provide a novel and convincing tool for distinguishing concerted and diradical mechanisms. Only the diradical pathways will be influenced by magnetic effects. For example, consider the simplified diradical mechanism for endoperoxide thermolysis shown in Scheme 5. If thermolysis leads initially to a singlet diradical ^1D, this species may undergo either ISC to ^3D (path b) or fragmentation of $^1O_2$ (path c). As in the case of radical pairs, magnetic fields may influence reactions of diradicals via a Δg effect or via a hyperfine effect. The Δg effect will increase the rate of step b relative to step c, thereby producing ^3D with greater efficiency. Hence, a higher yield of $^3O_2$ and a lower yield of $^1O_2$ is expected when endoperoxides (which decompose via diradicals) are thermolyzed

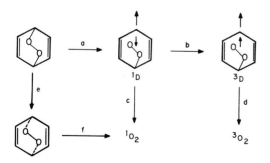

Scheme 5.

in a magnetic field. Furthermore, no effect of an external magnetic field is expected if a concerted decomposition occurs. Figure 21 shows a plot of the $^1O_2$ yield versus $\underline{H}$ for a 1,4-endoperoxide ($\underline{1}$) that undergoes concerted thermolysis and a 9,10-endoperoxide ($\underline{2}$) that undergoes thermolysis via a diradical.[55] It is extremely gratifying to find that there is no magnetic field effect on the $^1O_2$ yield for thermolysis of $\underline{1}$, but that a striking <u>decrease</u> in the yield of $^1O_2$ is observed for $\underline{2}$ for variation of field strength in the range 9000-15,000 Gauss.

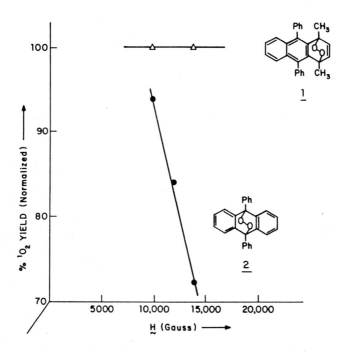

Fig. 21.   Yield of trappable $^1O_2$ from the thermolysis of some endoperoxides as a function of applied magnetic field.

A spectacular prediction can be made concerning magnetic isotope effects on the thermolysis of endoperoxides. If a diradical pathway (Scheme 5) is followed, diradicals possessing ^{17}O atoms will have a higher probability of following path b than diradicals possessing only ^{16}O or ^{18}O atoms, because ^{17}O is a magnetic isotope but ^{16}O and ^{18}O are non-magnetic isotopes. Experimentally, this means that endo-peroxide molecules which contain ^{17}O will produce $^1O_2$ <u>less</u> efficiently and $^3O_2$ more efficiently. Thus if a selective and efficient trap of $^1O_2$ is present during reaction, the "untrappable" molecular oxygen will be enriched in ^{17}O!

In the case of ^{17}O enrichment, when the representative point reaches the surface crossing (corresponding approximately to the diradical structure, Figure

22), the ^{17}O hyperfine interaction provides a mechanism for transition from the initial singlet surface to the triplet surface. No such mechanism exists for diradical structures possessing only ^{16}O and/or ^{18}O atoms. Thus, diradicals possessing at least one ^{17}O atom have the possibility of undergoing a more efficient ISC (and, therefore, produce $^3O_2$ more efficiently) than diradicals possessing only ^{16}O and/or ^{18}O.

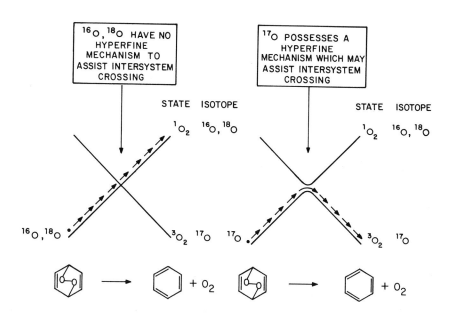

PREDICTION: (1) TRAPPED $^1O_2$ "ENRICHED" IN ^{16}O (AND ^{18}O)

(2) GASEOUS $^3O_2$ "ENRICHED" IN ^{17}O

Fig. 22.   Schematic surface diagram for the thermolysis of an endoperoxide by the diradical mechanism of Scheme 5. If the only mechanism for ISC were hyperfine coupling of odd electrons to oxygen nuclei (at the diradical structure) then only diradicals possessing ^{17}O atoms could undergo intersystem crossing.

To test the validity of these ideas, two types of measurements were made:[55] (a) the ^{17}O and (^{16}O + ^{18}O) content of untrappable molecular oxygen was analyzed by mass spectrometry, and (b) the yield of trapped $^1O_2$ was evaluated by quantitative determination of the amount of reacted acceptor when DPA-$^{16}O_2$, DPA-$^{18}O_2$ or

DPA-$^{17}O_2$ was employed (DPA is 9,10-diphenylanthracene, structure 2 in Figure 21). It was found that the yield of $^1O_2$ formation is smaller for ^{17}O than for ^{16}O or for ^{18}O containing DPA-O_2. Furthermore, it was found that both DPA-$^{16}O_2$ and DPA-$^{18}O_2$ produce the <u>same</u> yield of $^1O_2$, while DPA-$^{17}O_2$ produces less $^1O_2$. This result rules out a significant mass isotope effect as the basis for different quantum yields.

Since the amount of reacted trap is monitored in the yield measurements, they only provide an <u>indirect</u> test of the isotopic enrichment. A <u>direct</u> measurement involves determination of the isotopic composition of the untrappable molecular oxygen produced in the thermolysis of DPA-O_2. The results demonstrate that the untrappable molecular oxygen is indeed enriched in ^{17}O relative to the control sample.[55] Finally, endoperoxides which undergo concerted thermolysis were found not to produce ^{17}O molecular oxygen in equivalent experiments.

(F) <u>^{17}O Isotopic enrichment in the autoxidation of ethylbenzene</u>

Buchachenko and coworkers[56] have detected a considerable ^{17}O magnetic isotope effect during the autoxidation of ethylbenzene in the liquid phase. When a 60% (v/v) solution of ethylbenzene in benzene was oxidized at 50°C in a closed volume in the presence of oxygen and with dicyclohexyl peroxydicarbonate as radical initia the remaining oxygen after a consumption of 80-90% was selectively enriched in the isotope ^{17}O! Analysis by mass spectrometry revealed the residual oxygen to be enriched by 13% with the isotope ^{17}O, while the content of the ^{18}O isotope had decreased slightly by 0.22% (and consequently the content of ^{16}O had decreased by 12.78%). The selectivity for the only isotope with a nuclear spin unambiguously shows the operation of a magnetic isotope effect.

$$R\cdot + O_2 \longrightarrow R\text{-}O\text{-}O\cdot \tag{16}$$

$$2R\text{-}O\text{-}O\cdot \longrightarrow R\text{-}O\text{-}O\text{-}O\text{-}O\text{-}R \tag{17}$$

$$R\text{-}O\text{-}O\text{-}O\text{-}O\text{-}R \rightarrow R\text{-}O\cdot \; O\text{-}O \; \cdot O\text{-}R \rightarrow O_2 + 2R\text{-}O\cdot \tag{18}$$

In the process of autoxidation of hydrocarbons, two main stages involve radical pairs: the formation of the tetroxide (eq. 17) and its decomposition (eq. 18). In both reactions ^{17}O will enhance the efficiency of the formation of products (^{17}O increases the probability of the two peroxy radicals to recombine and indirectly increases the probability of the tetroxide to decompose to give molecular oxygen and alkoxy radicals). ^{17}O will be channelled through the reaction sequence (eq. 17,18) more efficiently and, therefore, will accumulate in the remaining atmosphere (with respect to ^{16}O and ^{17}O).

IX. THE ROLE OF MAGNETIC ISOTOPE EFFECTS IN STUDIES OF KINETIC ISOTOPE EFFECTS

Of the properties that distinguish the isotopes of a single chemical element, besides their differences in mass, a difference in nuclear magnetic moment (or spin) can also become important in the operation of a kinetic isotope effect. In the presentation of experimentally observed and mechanistically verified magnetic isotope effects, it becomes apparent that they can reach considerable magnitudes also (or especially) for heavy isotopes, in contrast to the diminishing isotopic mass effects under these circumstances.

Magnetic isotope effects are to be expected if radical intermediates or diradicals occur in the course of a chemical reaction. The possibility for the contribution of magnetic isotope effects to the overall kinetic isotope effects in these reactions, therefore, must not be neglected. Even small efficiencies of pair reactions of correlated radical pairs (or diradicals) can contribute significantly and to a similar degree as the conventional isotopic mass effect (for elements heavier than hydrogen).

The projection of the information obtained from the studies of kinetic isotope effects on the transition state structures is strictly valid only if the reaction has been shown conclusively that radicaloid species (radical pairs, diradicals...) are not involved. Criteria are available to characterize an isotope effect to be caused mainly by the nuclear magnetic moment or by the mass of the isotopic nucleus. These criteria should help to establish the validity of proposed reaction mechanisms and transition state structures.

X. MAGNETIC ISOTOPE EFFECTS AS A POSSIBLE MEANS TO ENRICH HEAVY ISOTOPES

Kinetic isotope effects due to isotopic mass are significant only for elements of low atomic weight (Table 1). Efficient separation of these isotopes is, therefore, very circumstantial with conventional means.[57] A new technique to achieve efficient enrichment involving even heavier isotopes has recently become available, however, in the form of the laser isotope separation, suitable for specific systems.[58]

The magnetic isotope effect presents a new possibility for isotopic enrichment that is not limited to the light elements. The nuclear property (or non-property) to have spin (or not) should produce isotope effects (and consequently separation efficiencies) of similar magnitude for light and heavy elements (in terms of the hyperfine mechanism on isotope effects). Indeed, enrichment studies with various tin isotopes have been reported recently.[59] Similar effects are to be expected for other elements, if suitable chemical reactions are available for a spin specific generation of radical pair (or diradical) intermediates.

XI. CONCLUSION

Isotope effects in chemical systems are usually associated with the difference of the isotopic mass of the nuclei. The inclusion of an isotopic effect due to the nuclear magnetic moment still is completely neglected in even the most recent treatments of reactions of isotopic molecules.

A change of the view held by (organic) chemists concerning the possibility of

influencing reaction rates by nuclear spins and magnetic fields in general is
now appropriate, since a dramatic influence of these interactions have been solidly
demonstrated.

Importantly, magnetic isotope and magnetic field effects provide the organic
chemist with a new tool to study reaction mechanisms involving radical pairs (or
diradicals) and to probe their reaction dynamics by the familiar means of analyzing
the compounds involved in the chemical reactions after the reaction has been carrie
out.

ACKNOWLEDGEMENTS

The authors thank the United States National Science Foundation, the United
States Department of Energy, and the Swiss National Science Foundation for their
generous support of this research.

REFERENCES

1. J.F. Duncan and G.B. Cook, Isotopes in Chemistry, Clarendon Press, Oxford,
    United Kingdom, 1968.
2. C. Collins and N. Bowman (eds.), Isotope Effects in Chemical Reactions, Amer
    can Chemical Society Monograph, No. 167, Van Nostrand Reinhold Co., New York
    1970.
3. E. Buncel and C.C. Lee (eds.), Isotopes in Organic Chemistry, preceeding
    volumes of this series, Elsevier, New York.
4. L. Melander, Isotope Effects on Reaction Rates, Ronald Press, New York, 1960
    L. Melander and W.H. Saunders, Jr., Reaction Rates of Isotopic Molecules,
    John Wiley, New York, 1980.
5. J. Biegeleisen and M. Goeppert-Mayer, J. Chem. Phys., 15 (1947) 261;
    J. Biegeleisen, J. Chem. Phys., 17 (1949) 675.
6. J. Biegeleisen and M. Wolfsberg, Adv. Chem. Phys. 1 (1958) 15; M. Wolfsberg,
    Acc. Chem. Res. 5 (1972) 225.
7. K.B. Wiberg, Chem. Rev. 55 (1955) 713; K.B. Wiberg, Physical Organic Chemistr
    John Wiley, New York, 1964, p. 273 and 351.
8. A.L. Buchachenko, E.M. Galimov, V.V. Brshov, G.A. Nikiforov and A.D. Pershin
    Dokl. Akad. Nauk, SSSR, 228 (1976) 379; Proceedings Russ. Acad. Sci. 228
    (1976) 379; A.L. Buchachenko, Z. Fiz. Khim. 51 (1977) 2461; Russ. J. Phys.
    Chem. 51 (1977) 1445.
9. N.J. Turro and B. Kraeutler, J. Am. Chem. Soc. 100 (1978) 7432; N.J. Turro
    and B. Kraeutler, Acc. Chem. Res. 13 (1980) 369; N.J. Turro, Pure Appl.

Chem. 53 (1981) 259.

10. R.Z. Sagdeev, T.V. Leshina, M. Kamkha, O.I. Belchenko, YuN. Molin and A.L. Rezvukhin, Chem. Phys. Lett. 48 (1977) 89; Yu N. Molin, R.Z. Sagdeev and K.M. Salikhov, Soviet Scientific Reviews, 1B (1979) 1, M.E. Vol'pin (ed.).

11. H.C. Urey, J. Chem. Soc. (1947) 569.

12. J. Biegeleisen, Science, 110 (1949) 14; ibid., 147 (1965) 463.

13. R.P. Bell, Rev. Chem. Soc. 3 (1974) 513.

14. H. Eyring, J. Chem. Phys. 3 (1935) 107; H. Eyring, Chem. Rev. 17 (1935) 65.

15. M.E. Schneider and M.J. Stern, J. Am. Chem. Soc. 94 (1972) 1517.

16. A. Streitwieser, Jr., W.B. Hollyhead, G. Sonnischsen, A.H. Putjaamatka, C.J. Chang and T.C. Kruger, J. Am. Chem. Soc. 93 (1971) 5096.

17. H.F. Koch and D.B. Dahlberg, J. Am. Chem. Soc. 102, (1980) 6102.

18. P.W. Atkins, Chem. Brit. (1976) 214; P.W. Atkins and T.P. Lambert, Ann. Rep. Chem. Soc. A (1975), 67.

19. S. Glasstone, K.J. Laidler and H. Eyring, Theory of Rate Processes, McGraw-Hill Book Co., New York, 1940, p. 298 ff.

20. J. Bargon, H. Fischer, and U. Johnson, Z. Naturforschg. 22a (1967) 1551; ibid., 22a (1967) 1556.

21. H.R. Ward and R.G. Lawler, J. Am. Chem. Soc. 89 (1967) 5518; R.G. Lawler J. Am. Chem. Soc. 89 (1967) 5519.

22. R. Kaptein, Adv. Free Rad. Chem. 5 (1975) 381.

23. G.L. Closs, Proc. Int. Cong. Pure Appl. Chem. 23rd, 4 (1971) 19.

24. R.G. Lawler and G.T. Evans, Ind. Chim. Belg., 36 (1971) 1087.

25. A.L. Buchachenko, Usp. Khim. 45 (1976) 375; Russ. Chem. Rev. 45 (1976) 375.

26. R.Z. Sagdeev, K.M. Salikhov and Yu N. Molin, Usp. Khim. 46 (1977) 569; Russ. Chem. Rev. 46 (1977) 297.

27. H.R. Ward, Acc. Chem. Rev. 5 (1972) 18.

28. R. Kaptein, in Chemically Induced Magnetic Polarization, L.T Muus (ed.), D. Reidel, Dodrecht, Netherlands, 1977, p. 1.

29. A.L. Buchachenko, Usp. Khim. 45 (1976) 761; Russ. Chem. Rev. 45 (1976) 761.

30. A.L. Buchachenko and F.M. Zhidomirov, Usp. Khim. 40 (1971) 801; Russ. Chem. Rev. 40 (1971) 801.

31. G.L. Closs and C.E. Doubleday, J. Am. Chem. Soc. 95 (1973) 2735; G.L. Closs in Chemically Induced Magnetic Polarization, L.T. Muus (ed.), D. Reidel, Dordrecht, Netherlands, 1977, p. 225.

32. R.G. Lawler, J. Cm. Chem. Soc. 102 (1980) 430.

33. N.J. Turro, M.-F. Chow, C.-J. Chung, G.C. Weed and B. Kraeutler, J. Am. Chem. Soc. 102 (1980) 4843.

34. B. Brocklehurst, Nature, 221 (1969) 921; P.W. Atkins and G.T. Evans, Mol. Phys. 29 (1075) 921.

160

35. R. Kaptein, Chem. Comm. (1971) 732.

36. P. S. Engel, J. Am. Chem. Soc. 92 (1970) 6074; W.K. Robbins and R.H. Eastman, J. Am. Chem. Soc. 92 (1970) 6076.

37. H. Langhals and H. Fischer, Chem. Ber. 111 (1978) 543; B. Blank, P.G. Mennitt and H. Fischer, Proc. Int. Congr. Pure Appl. Chem. 23rd, 4 (1971) 1.

38. H. Paul and H. Fischer, Helv. Chim. Acta, 56 (1973) 1575; J.E. Burnett and B. Milne, Trans. Farad. Soc. 67 (1971) 1587.

39. R.B. Bernstein, J. Phys. Chem. 56 (1952) 893; R.B. Bernstein, Science, 126 (1957) 119.

40. B. Kraeutler and N.J. Turro, Chem. Phys. Lett. 70 (1980) 266.

41. N.J.Turro, B. Kraeutler and D.R. Anderson, J. Am. Chem. Soc. 101 (1979) 7435.

42. G. Brunton, A.C. McBay and K.U. Ingold, J. Am. Chem. Soc. 99 (1977) 4447.

43. L. Sterna, D. Ronis, S. Wolfe and A. Pines, J. Chem. Phys. 73 (1980) 5493; L. Sterna and A. Pines, 1977 Ann. Rep. Lawrence Berkeley Laboratory, 1BL-7355.

44. N.J. Turro, D.R. Anderson and B.Kraeutler, Tetrahedron Lett. (1980) 3.

45. J.H. Fendler and E.J. Fendler, Catalysis in Micellar and Macromolecular System Academic Press, New York 1975.

46. N.J. Turro and W.R. Cherry, J. Am. Chem. Soc. 100 (1978) 7431.

47. K.L. Mittal (ed.), Micellization, Solubilization and Microemulsions, Plenum Press, New York 1976.

48. N.J. Turro, M. Graetzel, and A.M. Braun, Angew. Chem. 92 (1980) 712; Angew. Chem. Int. Ed., 19 (1980) 675.

49. J.K. Thomas, Chem. Rev. 80 (1980) 283.

50. N.J. Turro, Modern Molecular Photochemistry, Benjamin/Cummings, Menlo Park, California, 1978.

51. G.C. Weed, M.-F. Chow, C.-J. Chung, unpublished results, Columbia University.

52. B. Kraeutler and N.J.Turro, Chem. Phys. Lett. 70 (1980) 270; N.J. Turro, M.-F. Chow and B. Kraeutler, Chem. Phys. Lett. 73 (1980) 545.

53. N.J. Turro and J. Mattay, Tetrahedron Lett. 21 (1980) 1799; J. Am. Chem. Soc. 102 (1981), 4200.

54. N.J. Turro, M.-F. Chow and J. Rigaudy, J. Am. Chem. Soc. 101 (1979) 1300.

55. N.J. Turro and M.-F. Chow, J. Am. Chem. Soc. 102 (1980) 1190.

56. V.A. Belyakov, V.I. Mal'tsev, E.M. Galimov and A.L. Buchachenko, Dokl. Acad. Nauk SSSR, 243 (1978) 924; Proc. Russ. Acad. Sci 243 (1978) 561.

57. G.E. Dunn in Isotope Effects in Organic Chemistry, vol. 3, E. Buncel and C.C. Lee (eds.), Elsevier, New York 1977, p.1.

58. R.V. Ambartzumian and V.S. Ketokhov, Acc. Chem. Res. 10 (1977) 61.

59. A.V. Podoplelpv, T.V. Leshina, R.Z. Sagdeev, Yu N. Molin, and V.I. Gol'danskii Pis'mo, Zh. Eksp. Teor. Fiz. 29 (1979) 419.

CHAPTER 4

BOND ORDER METHODS FOR CALCULATING ISOTOPE EFFECTS IN ORGANIC
REACTIONS

LESLIE B. SIMS
Department of Chemistry, North Carolina State University,
Raleigh, North Carolina (USA) 27695-8204

DAVID E. LEWIS
Department of Chemistry, Baylor University, Waco, Texas (USA)
76798

CONTENTS

I. INTRODUCTION

The term <u>isotope</u> <u>effect</u> refers to the difference in a chemical or physical property between chemical species (stable atoms, molecules, ions, etc.; reactive intermediates, free radicals, adducts, etc.; or species with no significant lifetime such as collision complexes, transition states, etc.) which differ only in their isotopic composition. By convention, changes in symmetry resulting from isotopic substitution are not included as isotope effects within this definition.

Substitution of one isotope of an atom for another is the simplest and chemically most similar substitution that can be made in a chemical species. The resulting isotope effect reflects not only the change in relative mass (which, except for the isotopes of hydrogen, is quite small), but also the nature of the bonding to, and the chemical environment of, the labeled atom. Isotope effects are often small, of the order of a few percent of the measured property value, the exception being substitution of hydrogen by deuterium or tritium at a position where bond making or breaking to the labeled atom occurs during the reaction, which leads to a <u>primary</u> isotope effect; labeling at positions where bonds are neither broken or formed can lead to <u>secondary</u> isotope effects <u>if</u> changes in bond strength to the labeled atom occur. Labeling at positions remote from major bonding changes where no changes in bond strength (and hence no changes in force constants) occur results in insignificant isotope effects. Fortunately, modern analytical techniques allow determination of isotopically-induced changes in properties of chemical species accurately and simply,[1-5] even when these are of quite small magnitude.

Although isotope effects reflect the nature of bonding only in the vicinity of the labeled atom, a successive labeling approach[6] - determination of isotope effects for labeling at different positions in successive experiments - can, in principle, reveal the nature of and, for kinetic isotope effects, the timing of bonding changes during a reaction. Determination of isotope effects may sometimes be accomplished on materials labeled only by natural abundance of the isotopes, but generally synthesis of starting materials labeled in specific locations is necessary. In practice, the experiments are often demanding and the interpretation of the results may not be straightforward. A combined experimental-theoretical effort has been advocated[7] as especially promising for isotope effect research. In such an approach, isotope effects are selected for determination at positions where the syntheses are practical and which, from experience, are expected to yield especially significant information. Subsequently, isotope effects are calculated on the basis of various models of the reacting system and compared with experiment; those models giving satisfactory agreement with experiment are thus identified for further study. The calculations can be extended to labeling at other positions, thus serving to indicate which isotope effects would yield significant additional information regarding the bonding and mechanism. Such an approach is practical because of the enormous advances in the facility with which theoretical calculations can be made and provides greatly enhanced abilities for interpreting isotope effects and for extracting bonding and mechanistic information from measured values. The process tends to be iterative, employing successively more sophisticated models and theoretical methods as more detailed information is obtained from or confirmed by experimental measurments.

The ability to perform calculations of isotope effects is thus an important research tool in organic chemistry, which requires three criteria to be met: the bases for preselected types of calculations must be learned; the means for performing the calculations, including the requisite computer programs, must be obtained; and the facility must be developed for translating the organic chemical concepts of structure and bonding and the physical organic concepts of selectivity and reactivity into the necessary mathematical format for input to the calculations, a process we call mathematical modeling of the reaction. Of these requirements, the first two present relatively little difficulty: the first is only a slight extension of the training of most organic chemists today, and the existence of several programs and program libraries[8] circumvents problems posed by the second. Mathematical modeling of reactions is less likely to be part of the formal training of organic chemists.

The aim of this article is to present the essence of one method for modeling organic reactions appropriate for isotope effect calculations within the framework of statistical thermodynamic theories. The method is termed the Bond Order method because it is closely related to the Bond Energy-Bond Order (BEBO) method developed by H.S. Johnston for calculating rates of hydrogen transfer reactions.[9] The method is based on the familiar concept of bond orders, which are the independent variables in the modeling scheme; other structural and bonding parameters necessary for the calculations are related to the bond orders by empirical or semi-empirical relations.

The Bond Order method represents a composite of work from several research groups over the past several years, and while we have given references to important sources in the original literature, no attempt has been made to credit all those whose work has contributed to this method of calculation, or to present a chronologically or otherwise complete review of methods for calculating isotope effects. The reader is directed to several[1,10-17] extensive and complete references. Our goal is to present an account of the Bond Order method which has been useful in interpreting isotope effects, particularly kinetic isotope effects, and relatively easily learned by chemists with no prior experience in performing calculations. In addition, we hope that the applications discussed will render it useful to our colleagues who have active research programs in calculations of isotope effects. The earlier volumes in this series[10] serve as extremely useful background material for chemists entering the field of isotope effects, provided that they are adequately acquainted with the basic relationships between equilibria and rates, mechanisms, and the theoretical bases of kinetic theories; this information is particularly important as background for calculations of isotope effects. We have, therefore, included references to some standard works in these areas in appropriate places in the text; those new to this area may wish to review the basic material before reading this account.

II. STATISTICAL THERMODYNAMIC EXPRESSIONS FOR ISOTOPE EFFECTS

For a chemical reaction which converts reactants R_i into products P_j and which proceeds through a transition state TS, statistical thermodynamics provides relations between the equilibrium constant K and the partition functions and energy difference between reactants and products[18] and between the rate constant k and the partition functions and energy difference between reactants and transition state.[19]

If one or more atoms in one of the reactants, say R', is substituted by another (usually heavier) isotope of the same element, an isotopic isomer *R' is formed which will react with equilibrium constant *K and rate constant *k. The equilibrium isotope effect (EIE) and the kinetic isotope effect (KIE) are defined as the ratios K/*K and k/*k, respecively. Substitution of the statistical mechanical expressions for the equilibrium and rate constants yields the following expressions for the isotope effects[1,20] (see Appendix for an extended devlopment of eqs. 1-5):

$$EIE = K/*K = (Q_{P'}/Q_{*P'})/(Q_{R'}/Q_{*R'}) \tag{1a}$$

$$KIE = k/*k = (Q_{TS}/Q_{*TS})/(Q_{R'}/Q_{*R'}) \tag{1b}$$

where the Q's are the molecular partition functions evaluated relative to the local potential energy extrema for R, P and TS. Thus, calculation of either kinetic or equilibrium isotope effects involves evaluating the ratio of isotopic partition functions Q/*Q for reactant isotopic isomers and either the transition state or product isotopic isomers, respectively.

The partition functions are generally evaluated within the rigid rotor, harmonic oscillator approximations, which yield[1,20]

$$(Q/*Q)_{trans} \cdot (Q/*Q)_{rot} = (M/*M)^{3/2} \cdot \prod_{i=1}^{n_{rot}} (I_i/*I_i)^{1/2} = mmi \tag{2a}$$

$$(Q/*Q)_{vib} = \{\exp(-\sum_{i=1}^{n_{vib}} \Delta u_i/2)\} \cdot \{\prod_{i=1}^{n_{vib}} (1-e^{-*u_i})/(1-e^{-u_i})\} = zpe \cdot exc \tag{2b}$$

$$(Q/*Q)_{elect} = 1 \tag{2c}$$

In eq. (2), M is the molecular mass, I_i the ith moment of inertia, $u_i = h\nu_i/kT$ (where ν_i is the frequency of the ith normal vibrational mode), $\Delta u_i = u_i - {}^*u_i$, n_{rot} is the number of rotational degrees of freedom (2 for linear and 3 for non-linear species), and n_{vib} is the number of genuine vibrational degrees of freedom ($3N-5$ for linear and $3N-6$ for non-linear N-atom molecules; $3N_{TS}-6$ for linear and $3N_{TS}-7$ for non-linear N_{TS}-atom transition states since the reaction coordinate frequency has been removed).

The translational and rotational contributions to the partition function ratio we term the mass-moment of inertia factor, mmi. The vibrational term yields two contributions, one arising from the zero-point energy, zpe, and one arising from population of excited vibrational states, exc. Note that substitution of atoms in a molecule by other isotopes of the same element does not change the electronic partition function.[21] The symmetry numbers which normally are included in the rotational partition function have been deleted from eq. (2) since they do not represent a contribution which is attributable to differences in structure or reactivity of isotopic isomers.

Substituion of eq. (2) into eq. (1) leads to the primary expression for calculating isotope effects:

$$IE = MMI \cdot ZPE \cdot EXC \tag{3}$$

where
$$MMI = (mmi)_{final}/(mmi)_R,$$
$$ZPE = (zpe)_{final}/(zpe)_R,$$
$$EXC = (exc)_{final}/(exc)_R,$$

and the final state refers to P' in the case of EIE and to TS in the case of KIE.

Note that MMI (the mass-moment of inertia term) represents a factor based on structural and mass differences of the isotopic isomers, whereas both ZPE (the zero-point energy contribution) and EXC (the vibrational excitation term) represent vibrational contributions to the IE. Differences in vibration frequencies depend not only upon mass differences between isotopic isomers, but also upon changes in structure (bond distances and angles) and in bonding (represented by force constant changes between reactants and products or transition state). Note also that MMI is temperature-independent, whereas both ZPE and EXC are temperature-dependent.[22] Stern and Wolfsberg[23] have shown that isotope effects in the absence of force constant changes (no-force-constant-change, or nfcc, effects) are vanishingly small;

that is, mass and structural differences <u>alone</u> do not lead to measureable isotope effects. Generally, this is reflected in an MMI factor which is very near unity. A corollary of this is that isotope effects are not good probes for geometry changes, but are instead sensitive probes of bonding (force constant) changes during chemical reactions.[24]

In the case of kinetic isotope effects in particular, a variation of eq. (3) is commonly employed. This results from application of the Teller-Redlich product rule[25] (see also Appendix) to isotopic isomers, which yields the alternative expression for isotope effects:

$$EIE = VP \cdot ZPE \cdot EXC \tag{4a}$$

$$KIE = (u_L/*u_L) \cdot (VP \cdot ZPE \cdot EXC) = RXC \cdot VP \cdot ZPE \cdot EXC \tag{4b}$$

where $\quad VP = (vp)_{final}/(vp)_R$, and $\quad vp = \prod_{i=1}^{n_{vib}} (u_i/*u_i) \tag{5}$

Equation (4b) is commonly employed rather than eq. (3) for evaluating KIE because the former contains only the isotopic frequencies and does not require explicit structural data (except as input for the determination of vibration frequencies if they are calculated), and because it leads to a factorization of the kinetic isotope effect into a dynamic factor ($u_L/*u_L$, the reaction coordinate frequency contribution RXC, which is also the high-temperature limit of KIE) and a structural factor (VP·ZPE·EXC). This factorization will be seen later to be useful in interpreting KIE.

If the isotopic substitution is for any atom other than hydrogen (any <u>heavy</u> atom), Bigeleisen[26] has shown that the factors in eq. (4b) can be expanded and simplified to yield a heavy atom approximation:

$$KIE = (u_L/*u_L) \cdot \{1 + \sum_{i=1}^{3N-6} (G(u_i) \Delta u_i)_R - \sum_{i=1}^{3N_{TS}-7} (G(u_i) \Delta u_i)_{TS}\} \tag{6}$$

where $G(u_i) = 1/2 + 1/u_i + 1/\{exp(u_i) - 1\}$ and $\Delta u_i = u_i - *u_i$. Equation 6 has been useful in approximating magnitudes for and in qualitative interpretations of KIE.

The question of the validity of the assumptions inherent in the primary equation (1-4) is often raised. Although a decisive and quantitative estimate of the errors which are introduced by the assumptions is difficult to arrive at, it is important to point out that the statistical thermodynamics used assumes independent particles, which means that the equations are strictly valid only for ideal gases. Despite this, however, there are good reasons to believe that these assumptions do not introduce significant errors when applied to reactions occurring in the gas phase or in dilute solution in a non-interacting solvent.[27-29] The separability of energy levels and the assumption of rigid rotor and harmonic oscillator behavior can all be corrected for, and the corrections are almost always very small in the case of IE calculations.[20] The principal reason for this can be seen from the form of eq. (1): the IE is a quotient of isotopic partition function ratios, and errors introduced into partition functions by uncertainties in the parameters used for their evaluation, or by approximations, are usually propagated as much smaller errors in $Q/*Q$ than in Q itself because isotopic isomers have nearly identical properties and a similar error will therfore be introduced in both Q and $*Q$. Furthermore, errors that do not cancel in taking the ratio $Q/*Q$ are likely to cancel in taking the quotient $(Q/*Q)_{final}/(Q/*Q)_R$, with the result that errors propagated into the calculated value of the IE from the assumptions and approximations used in deriving eq. (3) are generally thought to be small.

III. SIMPLIFIED PROCEDURES FOR CALCULATIONS OF ISOTOPE EFFECTS

The basic equations (1-4) for calculating isotope effects require for their evaluation: (i) structural data - bond distances r and bond angles $\emptyset$; and (ii) vibration frequencies ν_i, or $u_i = h\nu_i/kT$, for isotopic reactants and isotopic products or transition states. Both sets of parameters for stable species (reactants and products) may be obtained by experimental measurement; such is not the case, however, for transition states.

(A) Structural Parameters

In the foregoing discussion it was pointed out that isotope effects are quite insensitive to changes in structural parameters alone unless structural changes also induce changes in bonding parameters (force constants). The practical effect of this is that idealized bond distances and angles may be used for reactants and products, and those for the transition state may be estimated in a number of ways - by analogy with stable species, by interpolation between reactant and product (or reactive intermediate) values, by approximation using empirical or semi-empirical relations, or by chemical intuition - without seriously affecting the calculated IE.

(B) Force Fields

The important parameters in determining the calculated IE are the vibration frequencies of isotopic reactants and isotopic products or transition states. Whereas these can be measured for reactants and products in favorable cases, they must be obtained by performing a vibrational analysis for the transition state. A vibrational analysis yields harmonic frequencies, whereas experimental frequencies are anharmonic. Spindel and co-workers[29] have shown that more accurate KIE are calculated if harmonic frequencies are used for both transition states and reactants, which requires either that anharmonic corrections be made to experimental reactant frequencies (which is not simple or well-understood for many molecules) or, as Spindel suggests, that experimental reactant frequencies be used to refine a force field which is then subsequently used as the basis for a vibrational analysis to obtain harmonic reactant frequencies. A similar result (calculation of more accurate IE by use of harmonic frequencies for all species) is inferred for EIE calculations, especially if complete experimental frequencies are unavailable for either reactants or products. Thus, calculations of IE generally require vibrational analyses of four species: reactant isotopic isomers and either product or transition state isotopic isomers.

The important parameters for IE calculations thus become the force constants employed in the vibrational analyses. A careful examination of the primary eqs. (1-3) and considerable experience reveals that isotope effects are determined primarily by the force constant changes between reactants and products (EIE), or between reactants and transition state (KIE), and are much less dependent upon the magnitude of force constants employed for the reactant (although force fields employed should reasonably reproduce known experimental frequencies – to within a few percent (<10%) for any given frequency and somewhat better when averaged over all frequencies). Shiner and co-workers[30,31] have advocated that force fields for reactants and products in particular, and even that for the transition state, should be considered adequate only if they produce fractionation factors (Q/*Q relative to that of a standard molecule; Shiner has used acetylene as a standard reference for H/D and $^{12}C/^{13}C$ or $^{12}C/^{14}C$ isotopes) consistent with those for molecules of closely related structure.

Force constants based on internal coordinates R_i (bond stretches, valence angle bends, torsions, linear bends and out-of-plane bends) are most easily related to structure and bonding parameters and hence are preferred for isotope effect calculations where the goal is to deduce structure and bonding changes during chemical reactions. Such force constants are elements of a general valence force field (GVFF) – or coefficients in a quadratic (harmonic) potential energy function:[32]

$$2V = \sum_i F_{ii}R_i^2 + 2\sum_{j>i} f_{ij}R_iR_j = \underline{R}^t\underline{F}\underline{R} \tag{7}$$

The last expression in eq. (7) is the standard matrix form of the quadratic expression.[33] The indices i and j run over all 3N-6 vibrational coordinates; hence the force constant matrix $\underline{F}$ is a (symmetric) square matrix of order n = 3N-6. (This assumes that no redundant coordinates are included. This point will be discussed later.) The total number of force constants in eq. (7) is n(n+1)/2; fortunately the IE for labeling at a given position is determined primarily by those force constants relating coordinates involving the isotopic atom, which reduces the number of critical parameters.

Interaction force constants, f_{ij} (off-diagonal elements in $\underline{F}$), are found for most molecules to be small compared to the corresponding valence force constants, F_{ii} and F_{jj} (diagonal elements in $\underline{F}$), and therefore the second term in eq. (7) is often small compared to the first term (which is referred to as the simple valence force field-SVFF). An SVFF yields calculated frequencies within 20% of experimental values for most molecules and within 10% for many molecules, although addition of some interaction force constants is usually necessary to produce frequencies in close agreement with experiment. We have previously suggested[7,34] that an SVFF is usually sufficient for calculating isotope effects (except that one or more interaction constants are usually needed to produce a physically reasonable normal mode corresponding to reaction coordinate motion with zero or imaginary frequency, as discussed in a later section), and that the resulting interpretation of bonding (i.e., force constant) changes between reactants and transition state (or products for EIE) is thereby simplified. However, fractionation factors calculated using a SVFF often agree less well with those for related molecules than the comparison of calculated and experimental frequencies would suggest.

Williams[35] has pointed out that bonding changes between reactants and transition state (or products) can be related to force constant changes only if no redundant coordinates are included in the potential function, eq. (7), and only if the same set of coordinates is employed for reactants, transition state, and/or products. This is not always possible, and even when it is, the choice of a coordinate set may not be unique. In these cases, Williams suggests the use of so-called "relaxed" force constants which are reciprocals of the diagonal elements of the compliance matrix.[36] While this method appears to circumvent many of the problems associated with the use of valence force fields, compliance constants are not generally available for most molecules whereas valence force constants are quite generally available for a large number of molecules, and are suitably transferable for use in isotope effect calculations. The problem of redundancies is a more serious one which will be discussed in a later section.

Most force fields available in the literature are the result of fitting frequencies calculated from the force field to experimental frequencies for several isotopic isomers. Since syntheses involving substitution of D for H are generally simpler than those resulting in substitution of heavy-atom isotopes, and because the frequency shifts for D-labeled compounds are much larger than for heavy-atom frequency shifts, the overwhelming preponderence of force constant data in the literature are weighted in favor of higher frequencies (H-C stretching and H-C-L deformation modes, where the ligand L may be H or a heavy-atom). The low frequencies of molecular systems, however, contribute disproportinately to many physical phenomena, including thermodynamic properties, particularly entropies, of chemical species, entropy changes in chemical reactions and chemical rate processes, and kinetic isotope effects (see ref. 37 for an example of the effect of low frequencies on chemical reaction rates). It is, therefore, important that force fields employed in KIE calculations provide reasonably accurate low frequencies for the molecular models, and one low frequency-weighted force field has been reported.[38e]

More work is necessary to assess the adequacy of force fields, but current information suggests the use of valence force constants, with the addition of interaction force constants (those expected on the basis of experience to be most important) successively until good fractionation factors are obtained, as the simplest and most reasonable choice of force fields for calculations of isotope effects.

(C) Simplified Models

Even if simplified structural parameters and force fields are employed, isotope effect calculations (especially for organic reactions) involve relatively large molecules for which vibrational analyses are both complicated and intensive of computer (cpu) time. As was pointed out earlier, IE calculations usually involve vibrational analyses for four molecular species: reactant isotopic isomers and either product (for EIE) or transition state (for KIE) isotopic isomers. Standard computer programs for performing a vibrational analysis require either one[39] or two[40] matrix diagonalizations for each of which cpu time is approximately $\propto n^2$ (the matrix to be diagonalized is of order n = 3N-6 for an N-atom molecule, or larger if redundant coordinates are used or needed in the vibrational analysis). Besides involving large amounts of computer time, vibrational analyses of even moderately large molecules (N > 10) can be complicated and formidable because of the large number of structural and bonding parameters needed.

Fortunately, isotope effects normally reflect structure and (primarily) bonding changes only in the region of the isotopically-substituted atom and are very nearly independent of structure and bonding changes in regions of the molecules remote from the labeled position. Stern and Wolfsberg[41] have shown that for reactions of acyclic molecules, simplified models in which large changes are made in portions of the molecule remote from the isotopically-labeled atom, including complete omission (cutoff) may be used for calculations of IE without significant loss of computational accuracy. Such cutoff models involve many fewer structure and bonding parameters so that they are easier to construct, the results are easier to interpret, and they are also conservative of computer time. The Stern-Wolfsberg rules for constructing cutoff models for IE calculations[41] are summarized below:

1. Similar models should be used for the reactant and transition state (or product for an EIE).
2. For a "highly proper" cutoff model, all atoms necessary to define internal coordinates to the labeled atom must be retained; the atomic masses and the relative geometries of all atoms involved in the internal coordinates must be the same in the complete and cutoff models, and all force constants which involve the labeled position must be present and have the same values in the complete and cutoff models. There are two exceptions: changes in mass of atoms two or more bonds removed from the labeled position (i.e., replacement of a CH_3 group by a "mass 15" atom if the C atom is two or more bonds removed from the labeled positon) are allowed; and torsional coordinates with very small force constant (corresponding to "free rotation"), if there is no change in the force constant between reactant and transition state (for KIE) or product (for EIE), may be excluded.
3. A "proper" cutoff model includes all changes in geometry (bond lengths and angles) and force constants between reactant and transition state identically as in the full model if these parameters involve the labeled atom; geometry and force constants not involving changes at the labeled position between reactant and transition state may be different in the cutoff and the full-model calculation.

In practice, "proper" calculational models are obtained if all atoms at least two bonds from the labeled atom are retained and if coordinates involving geometry or force constant changes at the labeled position in going from reactant to transition state are included in the same way and with the same force constant in the full and cutoff models (whether or not other coordinates - those $\underline{not}$ involving geometry or force constant changes at the labeled position - are the same in the full and cutoff models). The difference ($\{(k/*k)_{full} - (k/*k)_{cutoff}\}/(k/*k)_{full}$) between isotope effects calculated using such models and those using full models is generally less than about 1% for H/D isotope effects and usually no more than a few tenths of a percent for heavy-atom isotope effects for reaction systems involving acyclic molecules.

Stern and Wolfsberg warned that cutoff procedures may be suspect in instances where extraordinarily large interaction constants coupling internal coordinates are expected, where torsional force constants are expected not to be very small or to change between reactant and transition state, or where other factors suggest that vibrations involving an isotopically-substituted atom may be significantly affected by structural or bonding changes remote from the isotopic positon. They suggest testing cutoff models for each reaction system by comparing a sample full-model and cutoff calculation, then changing the force field and repeating both the full-model and cutoff calculation; or by adding on additional atoms and repeating the cutoff calculation. If the cutoff is a proper one, the ratio of full-model to cutoff model calculations should be the same when either a different force field is used in both models, or when additional atoms are added to the cutoff model.

Many organic reactions involve cyclic moieties, which inherently include coupling (and usually larger interaction force constants) between more remote portions of the molecule, and torsional force constants which are nearly always larger than for acyclic analogues. There is some reason to suspect, therefore, that models which involve cutting off ring substituents or cutting through cycles may provide less satisfactory computational models than cutoff models for acyclic systems. In order to test this possibility, we have carried out a large number of calculations of exchange equilibrium constants, K_{ex}, and the corresponding equilibrium isotope effect, $K_{ex}/*K_{ex}$, and of kinetic isotope effects, $k/*k$, resulting from isotopic labeling at the exchange position, in which cyclic moieties including or adjacent to the exchanging or reacting position are involved.[42]   The results for models of S_N2 substitutions of benzyl chloride have already been reported.[43]   Figure 1 gives some additional results for the exchange between HD and benzene (eq. 8, a nfcc process[23]), and Table 1 presents some results for the carbon-14 KIE for an S_N1 displacement reaction from benzyl chloride.

$$*C_6H_5-H + HD \xrightarrow{\quad K_{ex} \quad} *C_6H_5-D + H_2 \qquad (8)$$

$$EIE = K_{ex}/*K_{ex} = {}^{12}K_{ex}/{}^{14}K_{ex}$$

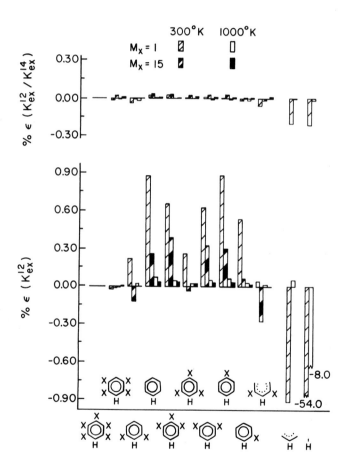

FIGURE 1.   Errors in EIE for reaction 8 of text, using various cutoff models.

The results in Figure 1 and in Table 1 are typical: cutting off substituents on a ring or cutting through the cycle (which requires also that the substituent at the last retained ring atom be cut off in order to satisfy the requirements for proper cutoff) results in less than ~1% error in H/D IE and only a few-tenths of a percent error for heavy-atom IE. Thus, it appears that the Stern-Wolfsberg cutoff rules apply with comparable validity for cyclic and for acyclic structures. We echo the suggestion of Stern and Wolfsberg that cutoff procedures should be validated for each system by a few trial calculations in which full model and cutoff model results are compared.

TABLE 1. Calculated Kinetic Isotope Effects ($^{12}k/^{14}k$) for Full and Cutoff Models of the Displacement of Cl^- from Benzyl Chloride According to an S_N1 Mechanism.

F_{*C-Ar} [a]	Full Model Ar-$^*CH_2Cl$ GVFF[b] or SVFF[c]	Cutoff Model Ar-$^*CH_2Cl$ GVFF[b]	VFF[c]
4.0	1.0791	1.0799	-----
4.5	-----	-----	1.0727
5.0	1.0684	1.0690	1.0656
7.0	1.0495	1.0503	1.0480
9.0	1.0330	1.0328	1.0325

[a] Stretching force constant for *C-Ar bond: reactant value = 4.5 md/A; *C-Cl bond: transition state value = 0.0; reactant value = 3.2 md/A.
[b] Generalized valence force field containing several interaction (off-diagonal) force constant elements in the potential energy matrix.
[c] Simple valence force field containing only diagonal force constant elements in the potential energy matrix.

(D) Summary of Steps Involved in Calculating Isotope Effects

    1. Choose appropriate models (cutoff or complete) for reactant and either product (for EIE) or transition state (for KIE).

    2. Calculate $(mmi)_{R'}$ and $(vp)_{R'}$ for the reactant model using experimentally derived or idealized values of r and ∅.

    3. Perform a vibrational analysis for the reactant model

using experimentally derived or idealized values of r, $\emptyset$, and F to obtain $(u_i)_R$, and $(*u_i)_{R'}$; calculate $(zpe)_{R'}$ and $(exc)_{R'}$.

4a. For an EIE calculation:
   repeat steps 2 and 3 for the product (P') model.

4b. For a KIE calculation:
   (i) estimate r_{TS} and $\emptyset_{TS}$ for the transition state model and either calculate $(mmi)_{TS}$ or:
   (ii) estimate F_{TS} for the transition state model and perform a vibrational analysis to obtain $(u_i)_{TS}$ and $(*u_i)_{TS}$; calculate $(zpe)_{TS}$, $(exc)_{TS}$ and $(vp)_{TS}$.

5. Calculate IE = MMI·EXC·ZPE or, for KIE = RXC·VP·EXC·ZPE, where the factors RXC, MMI, VP, ZPE, EXC are defined in eqs. (2) – (5).

Step 1 requires a straightforward application of the cutoff procedure.[41] Steps 2,3 and 4a require only appropriate structural (r, $\emptyset$) and bonding (F) parameters, either from appropriate tabulations[44,38] or by measurements. Given the insensitivity of isotope effects to structural parameters alone, the decision to use either experimental or idealized values of r and $\emptyset$ is not critical, particularly since the experimental values seldom differ from the idealized values by more than a few percent. For EIE, step 5 completes the calculation.

For the calculation of KIE, however, it is necessary to carry out step 4b rather than 4a, and this requires the estimation of transition state parameters (force constants in particular). It is to address this problem that the Bond Order method has been designed; the method is developed in the next section.

IV. BOND ORDER METHOD FOR CONSTRUCTING TRANSITION STATE MODELS

(A) The Concept of Bond Order and Transition State Diagrams

The concept of Bond Order was first introduced on a quantitative basis by Pauling[45] to rationalize bond distances in metals. The defining equation is (9), where r(n) is the length of a bond

$$r(n) = r(1) - 0.30\ln(n) \tag{9}$$

of order n and r(1) is the length of a single bond; the constant in eq. (9) was originally assigned the value 0.26 by Pauling. The Pauling relation (9) has since been shown to be a good and useful approximation for bond distances for many classes of chemical species, although modern structural data indicate the need for revision of the original value of the constant in eq. (9) if the relationship is assumed to apply generally. Table 2 shows a few representative comparisons between average experimental bond lengths and those calculated using values of 0.26 and 0.30 in eq. (9), the latter being a revised value which results in the best fit[46] of calculated to experimental bond lengths[44] for a very large number of common bonds. Bond orders (sometimes referred to as bond numbers) also are defined by most bonding theories, including Valence Bond and Molecular Orbital theories, and although the numerical values obtained are dependent on which theory is used to calculate them, the differences are small. The Pauling bond orders are especially useful because they are based on experimental results and because the values obtained agree, for simple molecules, with those derived from Valence Bond structures with which organic chemists are familiar. For those cases where resonance or Molecular Orbital descriptions are preferred, the derived bond orders are usually non-integral, but the Pauling bond orders and the bond orders from bonding theories are generally compatible. For example, common single bonds all have n=1; for olefinic, carbonyl, imino and sulfonyl groups, etc., n=2; for acetylenic and nitrile groups, n=3; the C-C bonds of benzene (r = 1.39) have a Pauling bond order of 1.65, which agrees well with the Molecular Orbital value of 1.67,[47] and with the Valence Bond value of 1.50 (based solely on the two Kekule resonance contributors). The small (10%) discrepancy between MO and VB bond orders for benzene is resolved by applying a more sophisticated VB approach.[48]

TABLE 2. Comparison of Experimental Bond Lengths and Those Calculated Using Equation (9) of Text with Constants of 0.30 and 0.26 (original Pauling value).

Bond	Order	Experimental[a]	Bond Lengths Calculated Using Value of Constant	
			0.30	0.26
C–C	1	1.537 ± 0.005	----	----
C=C	2	1.335 ± 0.005	1.329	1.357
C≡C	3	1.202 ± 0.005	1.207	1.251
C•••C (aromatic)	1.5	1.394 ± 0.005	1.415	1.432
C–O	1	1.426 ± 0.005	----	----
C=O	2	1.215 ± 0.005	1.218	1.246
O–O	1	1.48 ± 0.01	----	----
O=O	2	1.2742 ± 0.0002	1.272	1.300
N–N	1	1.451 ± 0.005	----	----
N=N	2	1.25	1.243	1.271
N≡N	3	1.0976 ± 0.0002	1.121	1.165
C–N	1	1.472 ± 0.005	----	----
C≡N	3	1.157 ± 0.005	1.142	1.186
Average Deviation from Experimental Value			0.009	0.029

[a]Reference 44.

Eq. (9) provides a transformation from bond distances to bond orders which is extremely useful in depicting chemical reactions. Fig. 2a is a conventional potential energy surface (PES) contour diagram for a concerted S_N2 reaction, eq. (10), as an example.

$$Nu:^- + RCH_2LG \longrightarrow NuCH_2R + LG:^- \qquad (10)$$

Figure 2b is the corresponding reaction coordinate diagram of energy vs. distance along the reaction coordinate (the dotted line in Fig. 2a) from reactants R through transition state TS to products P. The reaction coordinate for this reaction can be approximated as a linear combination of the C-Nucleophile (C-Nu) and C-Leaving group (C-LG) bond distances (for a more recent and complete description of the reaction coordinate diagram for S_N2

reactions see ref. 49). In principle, therefore, the distance between the reactant and product energy minima (i.e., the length of the abscissa) should approach infinity if the diagram is to reflect quantitatively the bonding in the transition state, since it is only at very large distances that the interaction between the reactants is sufficiently small to consider them to be independent particles. By using bond orders, however, the course of the reaction is simpler to represent since the bond order of the C-LG bond varies between 1.0 and 0.0 and that of the C-Nu bond from 0.0 to 1.0 as one proceeds from reactants to products.

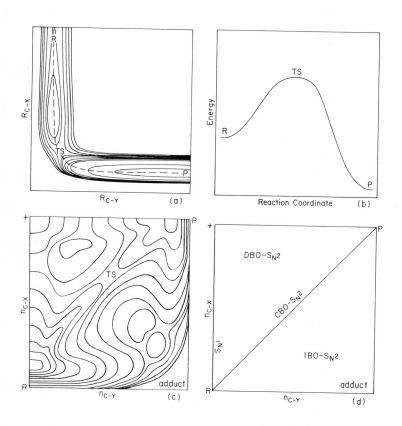

FIGURE 2. Different Representations of the Relationship Between Reactants, Products, and Transition State for Eq. (10) of Text.

R.A. More O'Ferrall[50] suggested that a useful alternative representation of the PES is obtained by using bond orders instead of bond distances. A More O'Ferrall contour diagram for an S_N2 reaction is shown in Fig. 2c; note that the reactant and product species occur in the lower left and upper right corners and that the reactive intermediates (the carbocation RCH_2^+, indicated by a + in the upper left of Fig. 2c, and the adduct $Nu-CH_2R-LG$, bearing the same net charge as the reactants) appear at the diagonally opposed upper left and lower right corners, respectively. The adduct may be considered as a covalent analogue of the solvent cluster proposed by Doering and Zeiss[51] to provide backside assistance in S_N1 reactions, with the nucleophile playing the role of the central solvent group. The presence of the carbocation and adduct species in Fig. 2c does not imply that these intermediates are formed in any particular S_N process, but rather that the energy of these possible species (relative to R and P) may affect the nature of the PES so as to cause a shift of the TS away from the central R$\longrightarrow$P diagonal (where TS for "synchronous" S_N2 processes should lie). Thus, changes in the reaction system which lower the energy of (i.e., stabilize) either the carbocation or the adduct (relative to R and P) will lead to TS with, respectively, increasing carbocation character (as is believed to occur for reactions of benzylic derivatives[52,53], vide infra) or increasing bond orders to both Nu and LG - i.e., more adduct character (as may occur, for example, in acyl transfer reactions and substitution at tetrahedral phosphorous centers[54], substitution reactions of square planar transition metal compounds[55], and in S_N reactions of alpha-nucleophiles[56,57]).

If Fig. 2c is thought of as a flexible, elastic surface connecting the species R, P, +, and adduct, More O'Ferrall[50] and Jencks[58] have shown that various rules for predicting structural and substituent effects on TS structure (such as the Hammond Postulate[59] and the rules of Thornton[60a] and Kurz[60b]) can be understood as the result of the effect of such changes upon the relative energies of the various species involved. Fry[61] first suggested that the infinite variations of Fig. 2c (as one varies the relative energies and identities of R, P, +, adduct) be projected onto a plane to produce a transition state diagram, as in Fig. 2d. In Fig. 2d, any line connecting R and P is a reaction coordinate for a possible PES, and some point along the reaction path represents the TS (minimum potential energy barrier between reactants and products). Thus, any point inside or on the axes of Fig. 2d may be a TS for some particular reaction system (R, P, solvent, etc.). Such a simplified diagram for S_N reactions was first proposed by Sims, et al.[62] and the mechanistic implications for TS structures of S_N reactions have been explored by Harris, et al.[53]

The diagonal connecting R and P in Fig. 2d corresponds to a concerted S_N2 process in which bond loss to the leaving group, Δn_{C-Lg}, is completely compensated for by bond formation to the nucleophile, Δn_{C-Nu} ($\Delta n = n_{TS} - n_R$), as in equation (11). Alternatively, structures along this diagonal (including the end

$$\Delta n_{C-LG} + \Delta n_{C-Nu} = 0 \quad \text{or} \quad n_{C-LG} + n_{C-Nu} = 1 \tag{11}$$

points R and P) conserve bond order in the sense that the sum of the bond orders n_{C-LG} and n_{C-Nu} is unity during the course of the reaction (R$\longrightarrow$TS$\longrightarrow$P) as implied by the second form of eq. (11); we refer to such processes as occurring with Constant Bond Order at the carbon at which substitution occurs: CBO_C. Likewise structures in the diagonally upper left half of Fig. 2d are said to be "loose" bonded structures or to have Decreased Bond Order at carbon: DBO_C; and those in the lower right triangular region are "tight" bonded structures which have Increased Bond Order, IBO_C. Similarly, TS below the secondary diagonal connecting the reactive intermediates are commonly referred to as "reactant-like" and those above as "product-like" structures.

Albery and Kreevoy[63] have suggested that the "product-like" character, P_x, and the "tight-loose" character, t_x, of the TS are useful descriptive variables. A convenient definition of these variables (a slight modification from that used by Albery and Kreevoy) in terms of the extents of bond breaking to the leaving group $(1-n_{C-LG})$ and bond making to the nucleophile (n_{C-Nu}) is given in (12). $P_x = 0$ corresponds to reactants; $P_x = 1$ to

$$P_x = (n_{C-Nu} - n_{C-Lg} + 1)/2 \tag{12a}$$
$$t_x = (n_{C-Nu} + n_{C-Lg} - 1) \tag{12b}$$

products. All TS below the secondary diagonal (connecting the reactive intermediates) have $0 < P_x < 0.5$ and are referred to as "reactant-like"; those TS above the secondary diagonal are characterized as "product-like", since for these $0.5 < P_x < 1.0$. Synchronous concerted TS along the principal diagonal connecting R and P have $t_x = 0$ (no change in bond order during the reaction);

those above the principal diagonal correspond to bond loss intermediate between a concerted TS ($t_x = 0$) and the carbocation intermediate ($t_x = -1$, complete loss of one bond) and are referred to as "loose" TS; for TS below the principal diagonal, bond formation to the nucleophile runs ahead of bond rupture to the leaving group, so that t_x is intermediate between the concerted value ($t_x = 0$) and that for a fully-formed Doering-Zeiss intermediate ($t_x = +1$, indicating one full bond more than in the reactants).

If eq. (12) is inverted, eq. (13) results, where the W's are

$$n_{C-Nu} = P_x + t_x/2 = W_m \qquad (13a)$$
$$(1 - n_{C-LG}) = P_x - t_x/2 = W_b \qquad (13b)$$

"weighting factors" (product-like characters) for the C-Nu bond being made (m) and for the C-LG bond being broken (b).[64]

Buddenbaum and Shiner[30] define alternative weighting factors in terms of the product-like character (eq. 12a) and a "concertedness" parameter a_x defined by eq. (14), which relates the extent of bond formation to the extent of bond rupture. The relationship between a_x and t_x is $a_x = \ln(P_x + t_x/2)/\ln(P_x - t_x/2)$.

$$n_{C-Nu} = (1 - n_{C-LG})^{a_x} \quad \text{or} \quad W_m = (W_b)^{a_x} \qquad (14)$$

In practice, isotope effects are calculated for several TS models spanning portions of the transition state diagram. This variety of TS models is constructed by either: a) varying values of n_{C-Nu} and n_{C-LG} over the range of interest and relating geometry and/or force constant parameters to the assigned bond orders by empirical relations derived from stable molecule data, as Sims et al[2,7,34] have advocated; or, b) assigning various values of P_x and t_x (or a_x) from the range of interest and determining W_m and W_b by analytical solution of eq. 13 (or by numerical solution of eqs. (12a) and (14)). Most TS parameters p_{TS} are then derived from reactant parameters p_R and product parameters p_P using eq. 15, employing W_m for parameters involving

$$p_{TS} = W \cdot p_P + (1-W) \cdot p_R \qquad (15)$$

the newly forming bond and W_b for parameters involving the rupturing bond. In addition to eq. 15, there are equations of constraint among the parameters which arise from geometrical and physical constraints of the molecular systems. Thus, not all parameters are linearly independent; the linearly independent parameters may be derived according to eq. (15), but the remaining parameters must be determined from the conditions of constraint, which are not always obvious. This approach has been advocated by Schowen[64] as providing TS parameters which are physically reasonable in that they are intermediate between reactant and product values.

Structural changes which result in different relative energies of the species R, P, +, and adduct, and which shift the TS along the reaction coordinate direction in Fig. 2d, are termed "parallel effects" (alternatively Hammond shifts); and those which shift the TS structure perpendicular to the reaction coordinate direction are termed "perpendicular effects" (alternatively, anti-Hammond shifts) by Thornton.[60]

(B) Bond Orders and Molecular Models

Transition state models employed in calculations of KIE are conveniently represented on transition state diagrams such as Fig. 2d for a nucleophilic displacement reaction, eq. (10). For such a process the bond orders of the rupturing bond to the leaving group, n_{C-Lg}, and that of the forming bond to the nucleophile, n_{C-Nu}, are the independent parameters of the models. In the Bond Order method of mathematical modeling, the structural and bonding parameters necessary for performing vibrational analyses and evaluating equations (1-4) for the isotope effects are related to bond orders by empirical expressions which are reliable for stable species. Pauling's rule (eq. 9) provides an example of an empirical relation which is applicable to a wide variety of bonds and is surprisingly accurate for stable molecules, as Table 2 indicates. The applicability of Pauling's rule to reactive intermediates or metastable or unstable species (such as transition states), and its accuracy when so applied, is not well-known, but we assume that it applies in such cases with validity comparable to stable species. Pauling's rule is used in the Bond Order method to calculate lengths of TS bonds of order n_{TS}, $r(n_{TS})$, using experimental or idealized values for the corresponding single bond length $r(1)$ in the reactant (for which $n_R = 1$). The remaining parameters (bond angles and force constants) must also be related to bond orders by empirical relations.

Bond angles for stable molecules do not appear to be related to bond orders by any simple relation comparable to Pauling's rule. Fortunately, changes in bond distances and angles during chemical reactions do not generally produce measureable isotope effects. Consequently, any reasonable estimates of bond angles should be sufficient for isotope effect calculations. For reactants (or products), experimental or idealized values of bond angles are used, as discussed previously. For TS models, bond angles are assumed to be intermediate between reactant and product values, or estimated from structural analogies.

The most commonly used correlation for stretching force constants is Badger's rule[65], eq. (16), relating the stretching force constants between atoms in rows i and j of the periodic

$$-\ln(F_{ij}) = (R_{ij} - a_{ij})/b_{ij} \tag{16}$$

table to the bond length. The empirical parameters a_{ij} and b_{ij} have been re-evaluated[66] using more recent data, and it is found that for H and elements in the next three rows of the periodic table, $0.24 \leq b_{ij} \leq 0.32$. Combining Badger's rule with Pauling's rule leads to eq. (17), where $F_R(n)$ is the stretching force constant for a bond of order n, and $F_R(1)$ is the corresponding

$$F_R(n) = n \cdot F_R(1) \tag{17}$$

stretching force constant for a single bond between the same two atoms. For reactants, experimental or idealized stretching force constants for single bonds are employed, and the Pauling-Badger relation (17) is used to calculate stretching force constants for bonds of order $n \neq 1$ in the reactant and TS models.

No relation of general validity which correlates bending force constants (for valence angle bending, out-of-plane bending or torsion) appears in the literature. By analogy to the Pauling-Badger rule for stretching force constants, Johnston[9] has proposed relation (18) for the bending force constant for valence angle subtended by bonds i-j and j-k, of order n_{ij} and n_{jk},

$$F_\emptyset(n_{ij}, n_{jk}) = n_{ij} \cdot n_{jk} \cdot F_\emptyset(1,1) \tag{18}$$

respectively; $F_\emptyset(1,1)$ is the corresponding bending force constant when both i-j and j-k are single bonds. Several modifications of this rule have been suggested[67], but no real justification for any of these rules has been provided, and none of them reproduces experimental trends in angle bending force constants. For example, HCC bending force constants decrease in the series ethane (n_{C-C} = 1), ethylene (n_{C-C} = 2), acetylene (n_{C-C} = 3), whereas eq. (18) and the other proposed relations predict a marked increase. To our knowledge, no relation which includes the effect of bond angle upon the angle bending force constant has been proposed, but simple and convincing arguments[68] suggest that the bending force constants should decrease as the bond angle increases, other things being equal, principally because of decreased non-bonded interactions. On the basis of an examination of literature values for bending force constants and extensive testing in calculations of isotope effects, we propose[34] eq. (19) as a reasonable approxi-

$$F_\emptyset(n_{ij},n_{jk},\emptyset) = (n_{ij} \cdot n_{jk})^X \cdot g(\emptyset) \cdot F_\emptyset(1,1,T_d) \qquad (19)$$

mation for force constants for bending of HCH, HCC and CCC valence angles. In eq. (19), $g(\emptyset)$ is a "geometry" or hybridization factor

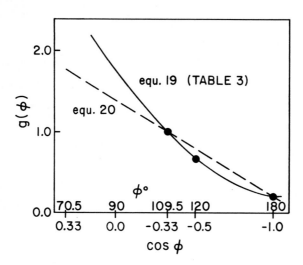

FIGURE 3. Geometry Factor of Equation (19) of Text (solid line) and the Analytical Function, Eq. (20).

when the valence angle has the value $\emptyset$ and $F(1,1,T_d)$ is the bending force constant when the subtending bonds are both single bonds ($n_{ij} = n_{jk} = 1$) and the valence angle is tetrahedral ($T_d = 109.47°$; $g(T_d) = 1.0$). Values of x of 1 and 1/2 have both been used. If there is no major bonding change at either bond subtending the angle in going from R to TS, x = 1/2 seems to reproduce bending force constants and H/D KIE very well; if either bond i-j or j-k is undergoing major bonding change, x = 1 seems to be a preferable value. The function $g(\emptyset)$ is plotted in Figure 3 as a solid line.

Table 3 compares the values of bending force constants determined from vibrational analyses of representative molecules assuming idealized angles[69] with the values calculated using eq. (19) and the values of $g(\emptyset)$ listed in the footnote of Table 3. The agreement is good in most cases.

TABLE 3. Comparison of Bending Force Constants Calculated from Equation (19) with those Determined from Vibrational Analyses (observed).

Angle	Value	Bending Force Constant	
		Observed[b]	Calculated[a]
H–C–C	109.5	0.65	----
H–C=C	120	0.60	0.61
H–C≡C	180	0.24	0.25
H–C∙∙∙C	120	0.52	0.56
H–C–H	109.5	0.55	----
H–C–H	120	0.35	0.37
C–C–C	109.5	1.0	----
C–C=C	120	1.1	0.94
C–C≡C	180	0.37	0.38
C∙∙∙C∙∙∙C	120	1.03	1.11
C–C∙∙∙C	120	0.75	0.86
C=C=C	180	0.40	0.44

[a]The "geometry" factors found to give the best results for HCC, HCH and CCC angles in a wide variety of molecules are $g(109.5) = 1.0$; $g(120) = 1/1.5$; $g(180) = 1/4.5$.
[b]Standard values for $F(1,1,109.5)$ were taken from reference 38c. Other values listed as observed are average values for a large number of molecules; see reference 38. Units are md Å/rad^2.

When eq. (19) is used to calculate TS force constants, secondary H/D kinetic isotope effects calculated for values of $\emptyset_{HCX} < T_d$ (as in S_N2 reactions, for example), with $g(\emptyset)$ extrapolated according to the solid line of Fig. 3, are consistently much smaller than those observed, indicating that the TS bending force constants, which primarily determine the value of the H/D KIE,[70] are too large. Apparently, values of $g(\emptyset)$ derived from Table 3 overestimate bending force constants for small $\emptyset$. If, as expected, the dependence of bending force constants upon $\emptyset$ is due largely to non-bonded interactions, then one expects $g(\emptyset) \propto \cos(\emptyset)$. (By the law of cosines $d^2_{XX} = R^2_{XC} + R^2_{CY} - 2R_{XC}R_{CY}\cos\emptyset$ for a valence angle $\emptyset_{XCY}$). Equation (20) is a revised equation for $g(\emptyset)$.

$$g(\emptyset) = 1.39 + 1.17 \cdot \cos(\emptyset) \tag{20}$$

Values of $g(\emptyset)$ from eq. (20) agree fairly well with those in the footnote of Table 3 for $\emptyset > T_d$, and the secondary H/D KIE resulting from the bending force constants derived from eqs. (19) and (20) are in reasonably good agreement with experimental values. Eq. (19) is plotted in Fig. 3 as a dotted line.

Saunders[1] has pointed out that eqs. (18) and (19) both predict that $F \longrightarrow 0$ as either $n_{ij} \longrightarrow 0$ or $n_{jk} \longrightarrow 0$, whereas very often angle bending force constants become (non-zero) out-of-plane bending constants, F_{OPLA}, upon rupturing one bond. He proposes the use of eq. (21) in such instances. This relation, like eq. (18), does not take into account the variation in bending force

$$F_\emptyset(n_{ij}, n_{jk}) = n_{ij} \cdot n_{jk}\{F_\emptyset(1,1) - F_{OPLA}\} + F_{OPLA} \tag{21}$$

constant with bond angle which we believe to be an important feature of any physically reasonable model. This feature is retained in eq. (22) which, like eq. (21), provides a non-zero

$$F_\emptyset(n_{ij}, n_{jk}, \emptyset) = (n_{ij} \cdot n_{jk})^X g(\emptyset)\{F_\emptyset(1,1,T_d) - F_{OPLA}\} + F_{OPLA} \tag{22}$$

value of $F_\emptyset$ when either bond is ruptured. Reasonable values of $F_\emptyset$ are obtained using eq. (22) with $g(\emptyset)$ in eq. (20).

Out-of-plane bending force constants for stable molecules are consistently close to the value in eq. (23), especially if the

$$F_{OPLA} = 0.20 \text{ md A/rad}^2 \tag{23}$$

central atom is carbon. Generally it is assumed that OPLA force constants do not change significantly in the transformation from R to TS. The assumption is not critical since it is generally observed that KIE are insensitive to OPLA force constants.

According to molecular orbital theory, single and triple bonds should exhibit essentially free rotation, whereas double bonds should be rigid. Torsional force constants should therefore vary with the bond order, decreasing from a maximum value for double bonds ($\sim$0.5 md A/rad^2 for alkenes), intermediate values for aromatic or conjugated systems ($\sim$0.3 md A/rad^2 for arenes), to very small values for single or triple bonds (0.072 md A/rad^2 for ethane is a representative value for alkanes). In a TS, the extent of double bond formation is limited by the alignment of the developing p orbitals at either end of the bond, which in turn is related to the state of hybridization of the terminal atoms of the bond. The hybridization can be measured by the angle $\emptyset_i$ between bond i-j and the non-reacting groups attached to C_i (i.e., the groups for which no primary bonding change is occurring during the reaction), with a similar definition for $\emptyset_j$ (values of $\emptyset$ are 109.5° for sp^3 hybridization; 120° for sp^2; 180° for sp). Equation (24) is an empirical equation for the torsional force

$$F_{tors} = n_{ij}{}^4 h_i h_j (F^0)_{tors} \qquad (24)$$

constant about a C_i-C_j bond in terms of the bond order n_{ij} and the angles $\emptyset_i$ and $\emptyset_j$, which agrees well with values observed for stable molecules. In eq. (24), $h_i = \{(T_d \cdot sin\emptyset_i)/(\emptyset_i \cdot sinT_d)\}^2$ and $(F^0)_{tors} = 0.072$ md$\cdot$A/rad^2, the standard value for ethane.[40] Equation (24) is employed both for reactants (where generally idealized angles are used for $\emptyset$) and transition states (for which variations of $\emptyset$ in going from R to TS is part of the modeling scheme). In most instances, KIE are very insensitive to torsional force constants, so that even very simplified approximations for $\emptyset$ are sufficient in eq. (24).

One interesting feature of eqs. (19), (22) and (24) is that they involve bond angle explicitly, so that changes in bond angle during the reaction now produce measureable isotope effects because they induce changes in bending force constants; inherently this seems reasonable. But if one uses these relations, as we suggest, changes in bond angle must be considered carefully, since now there will be an effect upon force constants which results in a contribution to the calculated isotope effect.

Equations (9) and (17-24) provide a set of expressions for relating transition state structural and bonding parameters to those of the reactant in an empirical, but physically reasonable way. The relationships are based on currently accepted theories and provide good approximations to the available data for stable molecules and should, therefore, be good approximations for the transition state values. Perhaps it should be pointed out again that in the absence of force constant changes, structural changes alone do not significantly affect the values of calculated isotope effects; and that certain force constants (e.g. out-of-plane bending and torsional force constants) do not generally much affect the calculated values of isotope effects, and can thus be estimated or even omitted in certain cases.

(C) Reaction Coordinate Motion and Tunneling

As eqs. (4b) and (6) indicate, kinetic isotope effects have two contributing factors to their value: a reaction coordinate or dynamic contribution, $(u_L/{}^*u_L)$, which derives from the unstable (i.e., $u_L = 0$ or imaginary) vibrational mode corresponding to motion along the reaction coordinate; and a structural contribution, (VPE·ZPE·EXC), deriving from geometry and stable vibrational frequency differences between reactant and TS. The dynamic factor can best be understood by reference to Fig. 2b which shows that, typically, reactants R must surmount an energy barrier (whose height is roughly equal to the activation energy) in order to proceed along the reaction coordinate through the TS region, at the top of the barrier, to products P. Motion along this coordinate can be shown to be a normal coordinate of the system and hence separable.[19a] The frequency associated with this reaction coordinate mode is determined by the effective reduced mass $m_{r,L}$ (which is a function of the relative atomic masses and structural parameters of the TS) and $F_L = (\partial^2 V/\partial R_L^2)$, the force constant along the reaction coordinate. Fig. 2b indicates that F_L is zero (for a broad, flat barrier with no curvature at the TS) or negative (for a narrow, curved barrier.) Since $m_{r,L}$ is always positive, the resulting frequency, $u_L = (h/kT)(F_L/4\pi^2 m_{r,L})^{1/2}$, is either zero (for a broad, flat barrier) or imaginary (for a narrow, curved barrier.)

Quantum mechanical tunneling becomes increasingly likely as

the barrier becomes more narrow and curved and as the effective mass decreases; i.e., as the (imaginary) reaction coordinate frequency becomes larger. Tunneling has been invoked to explain unusually large deuterium isotope effects[71] and recently proposed as a significant contributing factor to the magnitudes and temperature dependences of carbon KIE in elimination reactions.[72]

Equations (4b) and (6) do not include a tunneling correction and provide what is known as the semi-classical KIE. Several approximate tunnel corrections have been proposed based on different models for the barrier. The simplest and oldest model by Wigner[73] results in a tunnel correction to the rate constant of a multiplicative factor shown in eq. (25). The ratio of terms as in

$$(Q_t)_{Wigner} = (1 - u_L^2/24) \tag{25}$$

eq. (25) for light and heavy isotopic TS provides the (multiplicative) Wigner tunneling correction factor, $(Q_t)_{Wigner}$ to the semiclassical KIE.

A commonly-employed correction is the Bell tunneling correction,[74-76] the first term of which, (eq. 26), is most often

$$(Q_t)_{Bell} = (|u_L|/2)/\sin(|u_L|/2) \tag{26}$$

used as a multiplicative factor to correct for Bell tunneling. Again, the ratio of terms as in eq. (26) for light and heavy isotopic TS provides the Bell first-order tunnel correction to the KIE. Saunders[71a] discusses higher-order Bell corrections to KIE.

The dynamic or reaction coordinate contribution to KIE, as well as corrections for tunneling, depend upon the magnitude of the reaction coordinate frequency, ν_L. This contribution to KIE has been discussed by several authors[1,5,11,13,15,71a,77]. The reaction coordinate frequency, as well as the stable vibration frequencies of the TS, results from solution of the vibrational secular equation (27):

$$|\underline{GF} - \lambda\underline{E}| = 0 \tag{27}$$

In Eq. (27), $\underline{G}$ is the inverse of the kinetic energy matrix (see Sec. III.B.) and $\lambda = 4\pi^2 \nu^2$. Upon removal of any redundant coordinates[32,78], the matrix algebra assures that[32]

$$\det \underline{GF} = \det\underline{G} \ \det\underline{F} = \prod_{i} \lambda_i \tag{28}$$

For a TS, one (and only one) frequency (ν_L) is zero or imaginary; hence λ_L is zero or negative, respectively. The last term in eq. (28) is thus zero or negative, and since only positive values of the kinetic energy are allowed (which assures that $\det\underline{G}$ is positive), the criterion that a molecular system correspond to a TS rather than a stable molecule is that

$$\det\underline{F}_{TS} \leq 0 \tag{29}$$

As stated earlier, our knowledge of the structure and bonding of the TS is usually only conjectural, so that a simple valence force field (SVFF, diagonal $\underline{F}$ matrix) is often employed in calculations of TS vibration frequencies. For such a force field, eq. (29) reduces to

$$\det\underline{F}_{TS} = \prod_{i} F_{ii} \leq 0 \tag{30}$$

and in the TS one (and only one) valence force constant - that for the bond being ruptured - must be zero (corresponding to no restoring force) or negative (repulsive force). Such a TS model is appropriate (but not necessarily correct), for example, for an S_N1 or an E1 reaction in which one bond ruptures in the rate-determining step; the stretching of that bond would correspond to the reaction coordinate R_L and, at the TS, F_L would become zero or negative.

It is likely that more than one bond changes in the rate-determining step. In an S_N2 reaction, for example, the C-Nu bond is forming concurrently with (not necessarily synchronously with - i.e., not necessarily at the same rate as) rupture of the C-LG bond. In an E2 reaction, at least four bonds may be changing concurrently. It is important not only that ν_L be zero or imaginary, but also that the corresponding reaction coordinate vibrational motion (or mode) be such that bonds and angles are changing in concert as appropriate for the assumed mechanism. This is usually assured by adding one or more interaction force constants (off-diagonal $\underline{F}$ matrix elements) which couple the vibrational motions together. For example, in an S_N2 reaction an appropriate reaction coordinate mode would assure (perhaps among other things) that the C-Nu bond shortens and the C-LG bond lengthens in the reaction coordinate motion. If these bond stretches are labeled as vibrational coordinates 1 and 2, then addition of a coupling force constant between the two bonds would yield a potential energy matrix of the form below. Notice that $\underline{F}$ is always a symmetric matrix. On expansion, eq. (29) yields eq. (31) for this case. Note that eq. (30) is a special case of eq. (31) for $f_{12} = 0$. Eq. (31) may be satisfied without requiring that one of the valence force constants vanish or become repulsive (negative); instead, both chemical bonds may have normal force constants (albeit weaker than those in reactants or products). It

$$\begin{bmatrix} F_{11} & f_{12} & \emptyset & \emptyset & \cdots \\ f_{12} & F_{22} & \emptyset & \emptyset & \cdots \\ \emptyset & \emptyset & F_{33} & \emptyset & \cdots \\ \emptyset & \emptyset & \emptyset & F_{44} & \cdots \\ \cdot & \cdot & \cdot & \cdot & \cdots \end{bmatrix}$$

$$\det\underline{F} = (F_{11}F_{22} - f_{12}^2)(\prod_{i=3}^{3N-6} F_{ii}) \leq 0$$

$$= (1 - f_{12}^2/F_{11}F_{22})(\prod_{i=1}^{3N-6} F_{ii}) \leq 0 \tag{31}$$

is the concurrent, coupled motions of the two bonds which allows the reaction system to become a TS and thus effect displacement; mathematically this is achieved by the first factor in eq. (31) vanishing. If eq. (31) is divided by the last (non-zero) factor a more convenient form is obtained:

$$\det\underline{F}/(\prod_i F_{ii}) = (1 - f_{12}^2/F_{11}F_{22}) \leq 0 \tag{32}$$

Further simplification results from defining an interaction coefficient a_{12} as in eq. (33):

$$f_{12} = a_{12}(F_{11}F_{22})^{1/2} \tag{33}$$

The criterion for having a TS, eq. (29), becomes for this case:

$$(1 - a_{12}^2) = D \leq 0 \tag{34}$$

D is referred to as the _barrier_ _curvature_ _parameter_. For $a_{12}^2 = 1.0$, $u_L = 0$; for $a_{12}^2 > 1.0$, u_L is imaginary. If a_{12} is positive, u_L corresponds to the asymmetric motion of the two bonds (one lengthening, one shortening); if a_{12} is negative, u_L corresponds to a symmetric mode where both bonds lengthen or shorten.

The above procedure leading to eq. (32) can now be generalized to include coupling between any number of pairs of coordinates (usually adjacent bond stretches, or a bond stretch and bending of adjacent angles, are coupled rather than coordinates involving atoms remote from one another), yielding eq. (35):

$$\det \underline{F}/(\prod_i F_{ii}) = p(a_{ij}) = D \leq 0 \tag{35}$$

The exact form of the polynomial $p(a_{ij})$ must be derived by expansion of the $\underline{F}$ matrix containing the desired interaction constants, as was done above for the S_N2 example in deriving eqs. (31)-(34). This is not a formidable task if, as is usual, only a few coordinates are coupled.

Since the reaction coordinate motion is caused by addition of the interaction constant(s), eq. (7) indicates that $f_{ij}R_iR_j$ or, utilizing eq. (33), that $a_{ij}R_iR_j$ must be negative in order that the potential energy V decrease as one proceeds along the reaction coordinate from the TS toward product P (see Fig. 2b). If R_i and R_j are both of the same sign (both increasing or decreasing as in a symmetric motion of the two coordinates), then a_{ij} must be negative; if R_i and R_j are of opposite sign (an asymmetric motion of the two coordinates), a_{ij} must be positive. The sign of the a_{ij}'s required by a presumed reaction coordinate motion are thus simple to assess if one realizes that the convention for positive displacements of internal coordinates corresponds to bond stretching (increase in bond length) and angle closing (decrease in valence, linear, out-of-plane or torsional angle).

It is not clear, however, that only adjacent coordinates, or that only a few coordinate pairs, should be coupled. A general solution to the reaction coordinate problem which allows selection of the appropriate interaction constants to generate a preselected reaction coordinate motion and associated preselected reaction coordinate frequency has been given by Buddenbaum and Yankwich.[79] Buddenbaum and Shiner[80] have used the method for modeling S_N2 reactions of methyl halides using three interaction constants. In general, the Buddenbaum-Yankwich method does not afford an analytical solution for the interaction force constants f_{ij}, and an iterative numerical procedure must be employed. The method has not been widely applied, but it deserves more attention and study.

Modification of the force field to produce one zero or imaginary frequency affects all frequencies of the TS species, so that it is not so simple a matter to determine the reaction coordinate contribution RXC = $u_L/*u_L$ (see eq. 4b) to the isotope effect. It is, nevertheless, of considerable interest to assess the dynamic contribution (RXC, which is more clearly related to reactivity) and the structural contribution (VP·EXC·ZPE, which is related to bonding considerations).

Introduction of an interaction constant coupling two coordinates will lower one and raise another frequency (in actual fact all frequencies may be affected unless the coordinates are normal coordinates, although usually two frequencies will primarily be affected; the argument is unaffected if this is taken properly into account and generalized to any coordinate system) for both isotopic isomers of the molecular species. The lowered isotopic frequencies contribute an increase to exc and a decrease to zpe; see eq. (2); the raised frequencies contribute just the opposite effect. The net result is that changes in exc and zpe factors tend to compensate to a large extent in the ratio of isotopic partition functions Q/*Q. If similar force fields are used for reactants and final species (products for EIE; TS for KIE), there is a further compensation in the isotope effect expressions in eq. (3); the smaller the interaction constants, the better the compensation. Note that MMI is determined by stuctural parameters and masses only and is unaffected by the choice of force field.

The same effect (a trade-off in terms of changes in ZPE and EXC) is evident in comparisons of simplified (cutoff)[41] and full-molecule models for isotope effect calculations. This effect is one reason why Simple Valence Force Fields are often found to be satisfactory for calculations of isotope effects (which depend primarily upon isotopic frequency shifts and much less on the absolute magnitude of the frequencies), but not for calculating frequencies of individual species to compare with experimental values.

Shiner[81] has suggested that the reaction coordinate contribution can be determined from a "pseudo"-equilibrium isotope effect (p-EIE) between reactants and a "reaction complex" CPX having the same structural and bonding parameters as the TS except that interaction constants used to generate the reaction coordinate motion and frequency are absent (i.e., set equal to zero), and the KIE calculated for the same species transformed into a TS by addition of appropriate interaction constants, by eq. (36):

$$RXC = (KIE)/(p-EIE) \tag{36}$$

This suggestion is an extrapolation from experience that compensating errors in the isotope effect factors when using simplified force fields or simplified models often insure calculational accuracy. The extrapolation seems reasonable, but it has not been thoroughly examined. Recently, Schowen and Huskey[82] examined the approximation in some model calculations of hydrogen transfer reactions. They found that the approximation is valid for proper as well as very simplified models involving only atoms involved directly in the reaction coordinate formulation. Herein lies the possible importance of Shiner's suggestion: one can obtain the factors VP, EXC, ZPE using a large (proper or highly-proper cutoff or full-molecule) model TS and explore the reaction coordinate contribution RXC by means of eq. (36) using a much smaller model. The isotope effect expected for a large-model calculation employing reaction coordinate i would then be:

$$(KIE)_i = (RXC)_i(VP \cdot EXC \cdot ZPE)_{\text{large-model}} \tag{37}$$

If confirmed for other systems, this method could greatly reduce the cost and calculational effort of exploring reaction coordinate contributions to isotope effects for given TS structures.

Reaction coordinate motion may be represented on a reaction diagram, although not as precisely as the TS structure. Fig. 4 shows a reaction diagram for an elimination reaction system. The vertical axis represents possible TS for E1 processes; the horizontal axis contains possible E1cb TS; and all points in the body of the diagram represent possible E2 TS with varying degrees of E1 or E1cb character.

The structure of a TS is given by the coordinates (n_{C-H}, n_{C-X}) and the parameters determined from these independent coordinates by relations such as those given above; the Elcb-like E2 TS at the intersection of the curves in Fig. 4 is an example. Reaction coordinate A (RC-A) in Fig. 4 represents an Elcb-like process in which the C-H bond order is changing along the RC at the TS while the C-X bond order is essentially constant. RC-B is an El-like mode in which the C-X bond is rupturing while the C-H bond is unchanged. RC-C is an E2-like mode in which both C-H and C-X bonds are changing significantly and concurrently. RC-D is (probably) an unrealistic reaction coordinate mode since the motion at the TS is not in the general direction from reactants toward products.

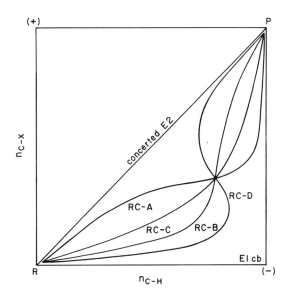

FIGURE 4. Reaction Coordinates for an Elcb-like E2 Reaction

Reaction coordinates A-D in Fig. 4 all apply to the same TS structure, and very different KIE may result from the models employing these various reaction coordinate modes. It should also be pointed out that such diagrams are limited as to what they can represent. In this case, the C-C bond is not represented at all, so no information concerning either the C-C bond order or the contribution of C-C bond contraction to the reaction coordinate motion is afforded by the diagram.

V. APPLICATIONS TO HYDROGEN TRANSFER REACTIONS

Hydrogen transfer reactions (eq. 38, where the prime indicates the hydrogen which is transferred, either as H^+, $H\cdot$ or

$$B + A-H' \longrightarrow B-H' + A \qquad\qquad (38)$$

H^-, are among the most common and most important reactions in chemistry. Because of the large relative difference in mass between protium, deuterium and tritium, primary kinetic isotope effects in hydrogen transfer reactions have the largest magnitude of any isotope effects ($^{H'}k/^{D'}k = 10$ have been observed fairly often). Even secondary H/D isotope effects are sizeable (up to ~50%; i.e., $^Hk/^Dk \sim 1.50$.) In contrast, primary heavy-atom (i.e., ^{14}C, ^{15}N, ^{18}O, ^{32}S, ^{37}Cl, ...) are generally less than 10% and only a few examples of a secondary heavy-atom isotope effect have been reported and these results, confirmed by model calculations, suggest magnitudes of less than 1% in favorable cases. Syntheses incorporating D in place of H are also generally simpler than those incorporating isotopes of heavy-atoms, and the starting materials (deuterated compounds) are generally inexpensive relative to compounds enriched in isotopes of heavy-atoms. The result is a robust literature on hydrogen isotope effects, including several comprehensive reviews.[83] Despite the ease of preparing materials and measuring relative rates, hydrogen isotope effects are still not completely understood, but model calculations have played a major role in our current understanding of these effects.

The first models of hydrogen transfer reactions were the three-centered models of Westheimer,[84a] Melander[84b] and Bigeleisen.[85] Three-center models are not "proper" in the Stern-Wolfsberg[41] sense, but for hydrogen isotope effects the models have been shown to be generally reliable.[1] These early models assumed a linear hydrogen transfer (in the TS, angle AH'B = 180°) with concerted changes in bonding such that the bonding loss $(1 - n_{A-H'})$ is exactly compensated by bond formation $(n_{B-H'})$ so that the bond order about the transferring atom remains constant, $n_{A-H'} + n_{B-H'} = 1$, during the transformation $R \longrightarrow TS \longrightarrow P$.

The earliest applications of this model[84,85] considered the effect of these bond changes only upon the (asymmetric and symmetric) stretching motions of the three-atom model, because the much higher stretching frequencies (~3000 cm^{-1}) would suffer a much larger frequency shift on isotopic substitution than the lower bending frequencies (~ 1000 cm^{-1}) and hence contribute disproportionately to KIE. Since the asymmetric mode becomes the reaction coordinate in these models of the reaction, the IE results primarily from the loss of asymmetric stretching frequency and the change in symmetric frequency in going from R$\longrightarrow$TS. These changes are manifested nearly entirely in the ZPE factor to the semi-classical IE (eq. 3,4) because substitution of D for H hardly changes the overall mass or moment-of-inertia of most molecules (and the differential change in the TS vs. that in R is even smaller) and hence MMI $\sim$ 1.0; and because few molecules are in excited vibrational states of these high frequencies at ordinary temperatures, and hence EXC $\sim$ 1.0. Furthermore, the symmetric stretching mode (the only genuine stretching vibration in the TS) is independent of the hydrogen mass if m_A and m_B (the masses of the cutoff models of A and B) are equal; and only moderately dependent upon m_H unless m_A and m_B have very different masses. Therefore, for many reactions, primary hydrogen isotope effects can be understood in terms of the reaction coordinate motion alone, and ZPE has often been taken as a measure of the reaction coordinate contribution to hydrogen isotope effects.

Westheimer[84a] and Melander[84b] demonstrated that the three-center stretching vibration model leads to KIE which are small for reactant-like or product-like TS (approaching unity in the limit as TS$\longrightarrow$R or TS$\longrightarrow$P in structure and bonding) and a maximum when $F_{A-H'} = F_{B-H'}$ in the TS. According to eq. (17), this would occur at $P_x = n_{A-H'} = 1 - n_{B-H'} = 0.5$ for a "symmetric" transfer reaction (one for which the reactant bond force constant, $F^o_{B-H'}$ and the product bond force constant, $F^o_{A-H'}$ are equal), the case in curve 1 in Figure 5, but at other more reactant-like ($0 \leq P_x \leq 0.5$) or product-like ($0.5 \leq P_x \leq 1.0$) TS if the reactant and product bonds have different force constants.

The Hammond postulate[59] suggests that TS$\longrightarrow$R as the reaction becomes highly exothermic and _vice versa_. It is also easy to demonstrate that a larger absolute maximum occurs (and for a more reactant-like TS) if the new (product) bond B-H' has a smaller force constant (generally this implies also a thermodynamically less stable bond) than the rupturing (reactant) bond A-H', as shown in Fig. 5, curve 2; curve 3 represents reactions in which bond A-H' is replaced by a stiffer (higher force constant) bond (generally a more stable bond also, as for an exothermic reaction). The predicted Westheimer maximum in KIE as the product-like character (P_x) of the TS increases has been demonstrated experimentally.[71a,86]

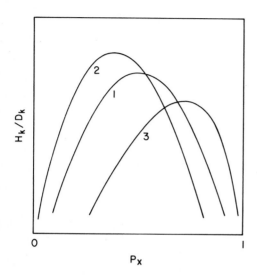

FIGURE 5. Hydrogen Transfer Kinetic Isotope Effects vs. P_x

The preferred linear arrangement and the assumption of constant total bonding about a transferring H are consistent with quantum mechanical considerations.[9,87] The other features and conclusions of the three-center models are consistent with more detailed proper or highly proper cutoff model calculations which retain the asymmetric stretch reaction coordinate but which take into account frequency shifts of all vibrational modes.[88]

Lewis[89] first suggested that in certain reactions, hydrogen may of necessity be transferred non-linearly and that different KIE may be exhibited by such systems. More O'Ferrall[90] has carried out model calculations which support the thesis that hydrogen isotope effects are smaller, other things being equal, as the angle of transfer decreases. Small KIE may also result from reactions for which $A_{H'}/A_{D'}$ (obtained from an Arrhenius plot of $\ln(^{H'}k/^{D'}k)$ vs. T^{-1}) is abnormally small (less than ~ 0.7, which also has been considered diagnostic of the importance of tunneling to the KIE[1,71a,22a]). Confusion exists about the relationship of the temperature-dependence of H/D KIE and the angle of transfer of the hydrogen. Kwart[22c] has maintained that "abnormal" T-dependence of H/D KIE implies non-linear hydrogen transfer. However, recent experimental work[22d] and model calculations [82] indicate that such a correlation probably does not exist. Several hydrogen transfer reactions have H/D KIE with surprisingly little T-dependence, and this is not well-understood. Stern and Wolfsberg[22a] and Yankwich et al[22b] have demonstrated by more elaborate models for several reaction systems that isotope effects may have various (some very peculiar) temperature profiles. The reactions examined by Kwart need to be better characterized mechanistically and more elaborate model calculations carried out in order to understand the KIE in these systems.

In contrast to primary hydrogen isotope effects, which exhibit the Westheimer maximum with increasing P_x, secondary hydrogen isotope effects should vary smoothly between 1.0 and the equilibrium value (EIE) as the transition state varies from reactant-like ($P_x \sim 0$) to product-like ($P_x \sim 1$).[91] Indeed, Schowen[64] suggests that secondary deuterium KIE may be used to infer P_x. However, anomalous secondary D-KIE (larger than EIE) have been observed by Kurz and Frieden[92a] and by Cleland, et al[92b] on hydride transfer reactions with NADH as hydride donor; in the same systems, large primary KIE are observed ($^Hk/^Dk = 5$-6.3). Secondary, alpha D-KIE are known to arise largely from changes in bending force constants to the isotopic atom on going from R to TS,[70] and it was suggested that coupling of bending vibrations of the secondary hydrogen atom into the reaction coordinate may give rise to anomalous secondary H/D KIE in hydrogen transfer reactions. It is expected that coupling of bending motions of the secondary hydrogen to the stretching vibrations of the transferring hydrogen will increase the ZPE (reaction coordinate) contribution to the secondary H/D KIE but reduce that contribution to the primary H/D KIE. The magnitude of the primary isotope effect may remain large if there is a significant tunneling contribution to the KIE, which is expected for many reactions.

Schowen and Huskey[64,82] have carried out extensive calculations of secondary and primary D-KIE for proper and simplified models of NADH-promoted hydride transfer reactions. The models retained the linear transfer and constant bond order about the transferring H of earlier models,[84,85] but more recent structural and bonding parameters were employed and reaction coordinate formulations included coupling of H-C-H' bends (H' is the transferring and H the secondary, non-transferring hydrogen atom) with the bond stretches of the rupturing C-H' and forming H'-C bonds. Anomalously large secondary H/D KIE (greater than EIE) and large primary H/D KIE resulted for (probably unrealistic) models with a very large loss of bonding ($n_{A-H'} + n_{B-H'} < 0.3$) or for models for which there were large tunneling contributions (corresponding to $\nu_L < 1000$ cm^{-1}) for a reaction coordinate involving substantial bending motion of the secondary hydrogen in addition to the dominant asymmetric stretching motion of the transferring H' atom. Such models exhibited significant deviations ($v > 1.0$) from the rule of the geometric mean[1] (The rule of the geometric mean, RGM, states in general that in a molecule with two or more chemically equivalent positions, the effect of isotopic substitution at one of these positions should be independent of the isotopic identity of the other positions; in this system, the RGM implies that $v = P_1/P_2 = S_1/S_2 = 1.0$, where $P_x = k(^xHC-H')/k(^xHC-D')$, for x = 1 or 2 for H or D, respectively, is a primary IE and $S_x = k(HC-^xH')/k(DC-^xH')$ is a secondary IE). Schowen and Huskey suggest that violation of the rule of the mean (which is experimentally determinable by measuring either both primary or both secondary KIE) is diagnostic of tunneling along a reaction coordinate involving both large primary and secondary hydrogen motion. The anomaly disappears if the transferring atom is a heavy-atom, so that secondary KIE in other substitution reactions should be less than the corresponding equilibrium isotope effect. More work is needed to completely explore the effects of reaction coordinate formulation on the magnitude of KIE in hydrogen transfer reactions.

VI. APPLICATIONS TO SUBSTITUTION REACTIONS

(A) Mechanistic Summary of S_N Reactions

Substitution reactions form one of the major classes of organic reactions. The general reaction type may be written as in eq. (39) involving a substrate with a leaving group (LG), X

$$-*\overset{|}{\underset{|}{C}}-X + Y \longrightarrow Y-*\overset{|}{\underset{|}{C}}- + X \qquad (39)$$

bonded to the central atom at which substitution is to occur (the *carbon), an attacking group Y, and other, non-reacting groups (NRG, indicated by dashed lines to *C in eq. (39)). The total bond order to *C is assumed to be 4 and may consist of four groups singly bonded, or fewer groups with multiple bonds to *C.

Substitution reactions are commonly classified[93] as electrophilic (S_E) or nucleophilic (S_N), depending upon whether Y is electron-deficient or electron-rich, respectively. We shall limit the discussion of this section to nucleophilic substitution reactions, in which Y is a neutral or negative ion species containing a lone pair of electrons. Nucleophilic substitutions are further classified[94] as unimolecular (S_N1) or bimolecular (S_N2) substitutions, depending upon the timing and extent of bonding change of the two bonds (*C-X and Y-*C) undergoing major changes in the rate-determining step of the reaction mechanism. In S_N1 processes, bond rupture of *C-X is rate-determining and occurs before attack of Y, thus producing a reactive intermediate carbocationic species, presumed to be planar. S_N2 processes occur with concerted *C-X bond rupture and Y-*C bond formation. The attack of Y is thought to be from the backside in most S_N2 cases, with formation of a trigonal, bipyramidal-like TS, although the angles X-*C-NRG may not be 90° as in a true trigonal bipyramid, and the bonding changes need not occur in a synchronous fashion – i.e., the extent of bond formation need not be equal to the extent of bond rupture. S_N1 and S_N2 processes lead to first- and second-order kinetics, and racemization and inversion of configuration, respectively. Many nucleophilic processes occur with mixed-order kinetics and some retention of stereochemical configuration; these are generally adequately accounted for by assuming that both paths occur competitively and simultaneously under many reaction conditions.

Winstein, et al[95] suggested that solvolysis reactions involve ion pairs, formed from the substrate RX in a series of physical-chemical processes involving formation of an ion pair in a solvent cage (R^+X^-, a "tight" ion pair), reorganization to ion pairs with intervening solvent ($R^+\|X^-$, "solvent-separated" ion pairs, where $\|$ represents the intervening solvent; a continuum of ion pairs in various degrees of intimacy and solvent intervention is proposed), to independent, solvated ions ($R^+ + X^-$, "solvated free ions"). According to this solvolysis scheme, substitution may occur by attack of a solvent molecule on any of the species depending upon reaction conditions and substrate, leading to a spectrum of rate-determining steps; the observed kinetics are pseudo-first order, of course. Sneen and Larsen[96] have suggested that all substitution reactions occur by a similar mechanism involving attack of the nucleophile upon one of the ion-pair species, leading in this case to observed kinetics intermediate between first and second order, with various stereochemical conse-quences. The ion pair mechanism does provide for a fundamental role of solvent as suggested by a host of experimental data and is able to accommodate the wide range of kinetics and stereochemistry observed in nucleophilic substitution reactions. However, there are difficulties in reconciling the ion pair mechanism with experimental observations in several systems,[97,98] and while it is believed to appropriately describe some cases, it is not accepted as universally applicable to substitution reactions, nor to provide a necessarily better explanation than the S_N1-S_N2 mechan-istic spectrum for the majority of cases where either scheme is consistent with observed kinetics. It is, moreover, possible to include solvent in a reasonable way within the S_N mechanistic scheme. Thus the application of the Bond Order method to nucelo-philic substitution reactions in this section will be presented within the framework of the S_N mechanistic scheme.

In a nucleophilic displacement reaction, the *C-X bond ruptures either before or in a concerted fashion with formation of the Y-*C bond. Since the leaving group X: departs with the pair of electrons from the rupturing bond, the transition structures occurring along the reaction coordinate are characterized by some charge development at the substitution center, *C (positive or negative depending upon the charge type of the reaction and the timing of bonding changes at the TS) . The energy requirement in such a reaction, and hence the rate, will be strongly influenced by the ability of the system to stabilize this charge. Because of steric crowding of the substituents and the nucleophile about *C, it is unlikely in an S_N2 reaction that intimate solvation at *C will play a major role in stabilizing the charge at *C, unless the solvent is also the nucleophile as in solvolysis reactions; solvent may play a decisive role in departure of the LG or approach of the nucleophile, however. Thus, except for solvolysis reactions, the groups directly bonded to the central atom in the substrate must account for most of the electron demand of *C accompanying departure of the LG in S_N reactions. If the groups initially present in the substrate are sufficiently good electron donating groups (EDG), it may be energetically favorable to rupture the *C-X bond in an S_N1 process, since the intermediate carbocation will be relatively stable; in this case, backside solvation by a solvent cluster as suggested by Doering and Zeiss[51] probably also contributes to stabilization.

An S_N2 process requires a close encounter, usually a backside attack, of a nucleophilic species and the stable substrate; initially these two closed-shell systems will repel (increase in energy), accompanied by a major energy loss as the *C-X bond begins to rupture, energy compensation of the new bonding inter-action of Y-*C, and stabilization of the positive charge by the electron density of the nucleophile. Differential solvation between reactant and transition states may also be a major con-tributor to the energy requirement. The external groups (nucleo-phile, solvent) and internal groups (non-reacting groups initially bonded to *C) therefore compete to supply the electron density demanded by the central atom and to facilitate departure of the LG. S_N1 and classical S_N2 processes are limiting cases in which the internal or external groups, respectively, exclusively provide electron density to compensate for that lost during departure of the LG. In general, both groups may contribute electron density, leading to a spectrum of S_N2 processes with TS of varying degrees of S_N1-like character. In Figure 2d (p. 180), S_N1 processes occur with TS lying along the ordinate, "synchronous" S_N2 processes occur with TS lying along the central diagonal connecting reactants and products, and TS in the upper triangular region of the figure correspond to the S_N1-like S_N2 processes in which the central atom carries a progressively greater formal charge as one progresses more toward the ordinate. The lower triangular region corresponds to processes with TS where bond formation runs ahead of bond rupture leading to "adduct-like" TS which are probably unrealistic for substitution about saturated carbon and will therefore not be discussed further in this section.

Some general factors affecting the mechanistic path follow from the above discussion: S_N1 processes show loss of stereochemical configuration, first-order kinetics, and a reactivity which increases with increasing substitution and substitution by EDG at the displacement center; S_N2 processes exhibit inversion of configuration, second-order kinetics and decreasing reactivity with increasing substitution, with pronounced dependence upon subtituent bulk but less upon electronic nature of the substituent. Highly polar solvents may aid in stabilizing the charged species and thus enhance an S_N1 mechanism; addition of salts enhances the ionizing power of the solvent and also favors an S_N1 process. Finally, the nature of the leaving group (especially ability to carry negative charge) and nucleophile affect the tendency of a system to react by an S_N1 or an S_N2 path.

Kinetic isotope effects observed for labeling at the displacement center (with ^{14}C), the non-reacting hydrogens attached to *C (with deuterium), and the leaving group (as appropriate; with ^{37}Cl, for example, for substitution reactions of alkyl chlorides) reflect the bonding changes occuring during the steps preceding and including the rate-determining stepof the mechanism; interpretation of observed values of KIE should lead to information regarding the timing and extent of such bonding changes. Leaving group KIE should reflect the extent of bond rupture to the leaving group at the TS; in practice, the values observed do not vary systematically or over a wide range for reactions for which very different extents of C-LG bond rupture are indicated by other evidence. For example, in the solvolysis of a series of para-substituted benzyl chlorides,[99b] the observed ^{37}Cl-KIE varied remarkably little in spite of widely different stabilities of the intermediate carbocations and hence the presumed extent of carbon-chlorine bond rupture and net positive charge buildup in the TS. It has been suggested[6,99] that "solvent assistance" of the LG leads to a "leveling" of the observed values and that differential solvation of the developing leaving moiety provides a source of bonding which compensates for bonding loss to *C to an extent which depends upon reaction conditions and system. In a sense the LG can be viewed as a "transfer" of LG from *C to solvent, and a transfer (bell-shaped) isotope effect behavior (refer to Sect. V) would be expected; since solvation bonds are generally much weaker than covalent bonds, the expected bell-shaped behavior of KIE (as the TS proceeds from reactant-like to product-like structure - a solvated leaving moiety in this instance - in response to changing reaction system or conditions) may not be realized, but a decrease in observed value would nonetheless be expected.

Table 4 gives the general ranges of kinetic isotope effects observed[100,101] in nucleophilic displacement reactions of para-substituted benzyl chlorides, as an example of a reaction system in which the mechanism may be varied between fairly wide ranges of the S_N spectrum by changes in subsituent, nucleophile, and solvent system. Calculations of KIE expected for "synchronous" S_N2 TS

TABLE 4.   Experimental Ranges of Kinetic Isotope Effects for S_N1 and S_N2 Reactions of para-substituted Benzyl Chlorides

Solvent System, Nucleophile and p-Substituent Favoring Mechanism	Experimental Ranges of Kinetic Isotope Effects		
	C-14	D_2	Cl-37
S_N1	1.01–1.03	1.20–1.30	1.008–1.010
S_N2	1.05–1.15	0.95–1.05	1.004–1.006

models (i.e., those for which extent of bond formation equals the extent of bond rupture) in this system have been communicated,[43] but the results from other S_N TS models have not been reported nor have the details of the models.   This system provides a particularly good example of the Bond Order method of modeling reaction systems and will be used as a detailed example of the method in the section to follow.   The remainder of the chapter will emphasize results obtained with the method for other reaction systems and reaction types, with a minimum of details included.

(B) S_N2 Models: The Benzyl Chloride System

1.   Reactant and TS Models for Substitutions of Benzyl Chlorides

The calculations of KIE for labeling at the central *C, the non-reacting hydrogens attached to *C, and the LG in the benzyl chloride system require, for proper cutoff, a minimum 7-atom reactant model and an 8-atom TS model; see Figure 6. Additional atoms would be required in the TS model if the nucleophile consists of more than one atom; for example, C-O would be the minimum nucleophile moiety for a TS model of an alkoxide ion attack on a benzyl chloride system.   In addition, one or more groups solvating the LG (each consisting of one or more atoms for proper cutoff) may be needed to account for solvent assistance of the LG and the resulting observed LG-KIE.

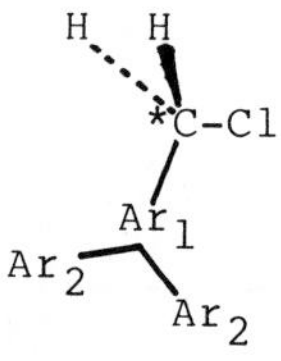

FIGURE 6a
Reactant Model

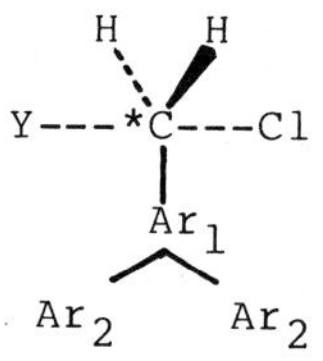

FIGURE 6b
Transition State Model

Fig. 6 reflects evidence from several sources that suggest that the geometry about *C is probably ~trigonal bipyramidal (with sp^2-like orbitals bonding the non-reacting hydrogens and the aryl group, and a developing p-orbital interacting simultaneously with the nucleophile and LG).   The aryl group is expected to be perpendicular to the Y-*C(Ar)-Cl plane if stabilization of the positive charge at *C is afforded by the aryl ring (by overlap with the aromatic π-system).

2.   Bond Orders of the Benzyl Chloride Models

In the reactant, bond orders of 1.0 are assumed for all bonds to *C.   Bond orders of 1.67 are assigned to the aromatic C-C bonds,[47] reflecting the aromatic stabilization.   Similar standard bond orders apply to the product.   The bond orders of the TS model, n(i-j), are the primary parameters of the modeling scheme, and will be assigned various values for different TS structures. The bond orders are summarized in Table 5.

If $\Delta N(*C) = N^{TS}(*C) - N^{R}(*C) = 0.0$ (i.e., if $N^{TS}(*C) = N^{R}(*C) = 4.0$), the TS is central, synchronous S_N2. If $\Delta N(*C) = -1$, the TS corresponds to an S_N1 process. For $-1 < \Delta N(*C) < 0$, the TS is S_N1-like S_N2. Note that $\Delta N(*C) = t_x$, the "tightness" parameter of Kreevoy and Albery (see Sect. IV.A.); the last two cases above correspond to "loose" TS in that terminology.

The hybridization of *C in the reactant is sp^3 and in the TS is approximately $sp^2 + p$; the non-reacting *C-H bonds thus approximate $*C(sp^3)-H$ and $*C(sp^2)-H$ in the reactant and TS, respectively. As a result, a small decrease in C-H bond length and a corresponding small increase in bond order and stretching force constant is expected; this conceivably could contribute to

TABLE 5. Bond Orders for Reactant, Product and TS Models for the Benzyl Chloride Substitution System.

	Bond Orders		
Bond	Reactant $CH_2(Ar)Cl$	TS $YCH_2(Ar)Cl$	Product $YCH_2(Ar)$
*C-Cl	1.0	$n^{TS}(*C-Cl)$	0.0
*C-H	1.0	$n^{TS}(*C-H)$	1.0
$*C-Ar_1$	1.0	$n^{TS}(*C-Ar)$	1.0
Ar_1-Ar_2	1.67^a	$n^{TS}(Ar_1-Ar_2)$	1.67^a
*C-Y	0.0	$n^{TS}(*C-Y)$	1.0
Total B.O.	$N^R(*C)=4.0$	$N^{TS}(*C)$	$N^P(*C)=4.0$
	$N^R(Ar_1)=4.33$	$N^{TS}(Ar_1)$	$N^P(Ar_1)=4.33$

[a] Ref. 47; values of $n(Ar_1-Ar_2)$ in excess of 1.0 and hence of $N(Ar_1)$ in excess of 4.0 represent aromatic stabilization of the aryl ring.

the observed alpha D_2-KIE. However, it is well known that secondary alpha deuterium KIE arise almost exclusively from changes in <u>bending</u> force constants,[70] and the small changes in bond order and stretching force constant of the *C-H bonds can be neglected to a first approximation and retained at the reactant values, eq. (40):

$$n^{TS}(*C-H) = n^{R}(*C-H) = 1.0 \tag{40}$$

Secondary heavy-atom isotope effects of significant magnitude have been reported in only a few cases,[102] and it is believed that such effects are generally vanishingly small. This implies that net bonding changes at heavy-atom centers not undergoing formal bonding change are very small. Thus for substitutions of benzyl chloride it can confidently be assumed that $\Delta N(Ar_1) \sim 0$ and that:

$$n(Ar_1-Ar_2) = \{4.33 - n(*C-Ar_1)\}/2 \tag{41}$$

In the absence of any interaction with the phenyl ring $\{n(*C-Ar_1) = 1.0\}$, localized charge $+q/e = \{1-n(*C-Cl)-n(*C-Y)\}$ develops at the central carbon. By conjugative interaction with the ring, a fraction F_{Ar} of this charge may be delocalized into the ring system. In valence bond terms, this is represented by increased importance of quinoid-type resonance structures in the TS; such structures contain a $*C=Ar_1$ double bond. Hence, conjugative stabilization by the aryl ring may be modeled by increasing the bond order of the $*C-Ar_1$ bond in proportion to the delocalization of charge into the ring, as given by eq. (42) with $0 < F_{Ar} < 1.0$.

$$n(*C-Ar_1) = 1.0 + F_{Ar}\{1 - n(*C-Cl) - n(*C-Y)\} \tag{42}$$

In order to investigate the possible importance of differential solvation of the LG, the TS model was expanded by addition of three solvating H-O moieties (cutoff models of a protic solvent) to complete a tetrahedral coordination sphere of the incipient chloride ion. The conformation of the groups about the *C--Cl bond was staggered. Some fraction F_S of the bond loss to the LG will be compensated for by bonding to each solvent moiety. The bond order between LG and solvent, $n(Cl--H)$ in this case, will thus be given by eq. (43), where $0 \leq F_S \leq 1$. The value of F_S can

$$n(Cl--H) = F_S\{1 - n(*C--Cl)\} \tag{43}$$

be estimated from Johnston's bond energy-bond order relation[9], eq. (44), where E_1 and E_n are the bond energies for a single bond and

$$\ln(E_n/E_1) = p \cdot \ln(n) \qquad (44)$$

for a bond of order n, respectively, and p is an empirical parameter determined by the identity of the atoms i and j forming the bond. For H-Cl, E_1 = 102 kcal/mole and p = 0.914[9].

The free energy of interaction between a gas-phase chloride ion and 1 - 3 water molecules is 6 - 12 kcal/mole.[103] Using these values as aproximations for E_n in eq. (44) yields bond order estimates for n(Cl--H) of 0.05 - 0.1, respectively, which serves to fix an approximate range for the parameter F_S of < 0.1.

The bond length r(Cl--H) was adjusted using Pauling's rule (eq. 9) and n(Cl--H) from eq. (43). Since F_S is small, the H-O bond order was assumed to be unity in the solvating group as in the hydroxylic solvent, and the bond length r(H-O) to be 0.97A as in water. The Cl--H-O moiety was assumed to be linear.

The _independent_ bond order parameters of the model are then the bond orders n(*C-Cl), n(*C-Y), the degree of interaction with the ring, F_{Ar}, and the extent of solvation, F_S; additional parameters will be needed to specify the reaction coordinate motion.

3. Geometries of the Benzyl Chloride Models

Geometrical parameters (bond distances and angles), as well as force constants, of the reactant and TS models must be related to standard values and to assumed and/or assigned bond orders of the models by empirical and/or semi-empirical relations (eqs. 9, 15, 17-22). Single bond lengths $r^o(i-j)$ were used for all single bonds i-j; for bonds of different order, lengths were adjusted from these single bond values by Pauling's rule (eq. 9). Standard values[44] employed were: C-H, 1.09; C-C, 1.54; C-Cl, 1.77; C-O, 1.43; Cl-H, 1.27; H-O, 0.97 (all values in Angstroms).

Bond angles in the reactant and product models were assumed to be tetrahedral (109.47°) about *C and planar trigonal (120°) about Ar_1. There are no rules which provide accurate estimates of how bond angles change as bond orders change, as Pauling's rule provides for bond distances. But since KIE are nearly independent of geometrical changes in going from reactant to TS, unless those changes induce changes in force constants as discussed earlier, physically reasonable estimates of TS valence angles are usually adequate for KIE calculations. In an S_N2 process, inversion of the NRG attached to *C occurs as Y displaces Cl. In an S_N1 process, these groups occupy trigonal positions in a plane ~perpendicular to the rupturing *C-Cl bond in the fully-formed carbocation. It is reasonable, therefore, to vary the H*CCl and Ar_1*CCl angles between the reactant value (109.47°) and the product values (90° for S_N1 processes; 180°-109.47° = 70.53° for S_N2 processes involving backside attack by Y), with the *C-Cl bond order. In the absence of compelling evidence about the dependence of valence angles upon bond order and in view of the fact that the calculations are insensitive to angles, a linear dependence is assumed, as shown in eqs. (45) and (46):

$$\emptyset_{i*CCl} = 70.53° + 38.94° n(*C-Cl) \qquad i=H, Ar_1 \qquad \text{for } S_N2 \qquad (45)$$
$$= 90.00° + 19.47° n(*C-Cl) \qquad i=H, Ar_1 \qquad \text{for } S_N1 \qquad (46)$$

Some researchers have assumed that the TS is planar trigonal ($\emptyset_{i*CCl} = 90°$, i = H, Ar_1) regardless of the value of n(*C--Cl). KIE calculated for models which differ only in the values of these angles are only slightly different for the two models.

4. Internal Coordinates for the Benzyl Chloride Models

Internal coordinates for the reactant model (Fig. 6a) were chosen by the rules of Decius[104] and are listed in Table 6.

The rules of Decius provide for an N-atom system a kinematically complete set of 3N-6 linearly-independent coordinates. For many systems which contain elements of symmetry, additional redundant (i.e., linearly-dependent) coordinates are required to assure that the potential and kinetic energy expressions reflect the full symmetry of the model. The required additional redundant coordinates are easily generated from the 3N-6 coordinates obtained from the Decius rules: Applying any of the symmetry operations appropriate to the model to any one of the 3N-6 internal coordinates produces a different internal coordinate. If the new coordinate is not among the linearly-independent set produced by the Decius rules, it must be included as an additional redundant coordinate.

For acyclic systems, redundancies occur only among angle bending coordinates. About any atom taking part in m $\geq$ 2 bonds there are m(m-1)/2 total angles, only (2m-3) of which are linearly-independent; the remaining (m-3)(m-2)/2 angles are related (by redundancy conditions) to the linearly-independent angles. Decius provides a method for choosing (2m-3) linearly-independent angles from among the total:

1. Select 3 atoms bonded to the central atom which are not coplanar with the central atom, and define three angles, each involving two of these atoms and the central atom.
2. For each additional atom, define two additional angles involving only this atom, the central atom and atoms tied together by angles already defined in previous steps.

In the reactant model, *C takes part in m=4 bonds; of the 6 angles about *C only 5 are linearly-independent vibrational coordinates. In preparing Table 6, H, Cl and Ar_1 were chosen to define $\emptyset(H*CCl)$, $\emptyset(H*CAr_1)$ and $\emptyset(Ar_1*CCl)$ by rule 1 above. Then, for the remaining atom H' bonded to *C, two additional angles $\emptyset(H'*CCl)$ and $\emptyset(H'*CAr_1)$ are defined. Note that in step 2, an arbitrary choice of which two angles to define was made, which resulted in the designation of the H*CH' angle as the linearly-dependent (redundant) angle in the group of six total angles about *C. Likewise, the initial choice of H,Cl, Ar_1 in step 1 (instead of H,Cl,H'; H',Cl,Ar_1; or H,H',Ar_1) was also arbitrary. Thus, the Decius method does not uniquely specify (2m-3) angles.

Table 6. Internal Coordinates for Reactant Model of Benzyl Chloride Substitution.[a]

Bond Stretches	Valence Angle Bends	Out-of-Plane Bends
R_1 = r(*C-Cl)	R_7 = $\emptyset(H*CCl)$	R_{16} = (*C-Ar)
R_2 = r(*C-H)	R_8 = $\emptyset(H'*CCl)$	
R_3 = r(*C-H')	R_9 = $\emptyset(Ar_1CCl)$	Torsion
R_4 = r(*C-Ar$_1$)	R_{10} = $\emptyset(H*CAR_1)$	R_{17} = (*C-Ar$_1$)
R_5 = r(Ar$_1$-Ar$_2$)	R_{11} = $\emptyset(H*CH')$[b]	
R_6 = r(Ar$_1$-Ar'$_2$)	R_{12} = $\emptyset(H'*CAr_1)$	
	R_{13} = $\emptyset(*CAr_1Ar_2)$	
	R_{14} = $\emptyset(*CAr_1Ar'_2)$	
	R_{15} = $\emptyset(Ar_2Ar_1Ar'_2)$[b]	

[a]The same set of coordinates is appropriate for the product model if Cl is replaced in the listing of the table by a one-atom cutoff model of Y.
[b]Non-essential (redundant) coordinates.

The minimum set of (3N-6) linearly-independent coordinates, plus any symmetrically-required coordinates, may be used as the coordinate set to perform a vibrational analysis (which minimizes the size of the vibrational problem and hence computer time), provided the additional (redundant) coordinates that may be defined are properly removed using standard techniques.[32,78] These techniques are, however, awkward to apply, and some researchers have simply deleted all redundant coordinates that were not required by symmetry considerations. If coordinates are deleted, it is generally desirable to delete symmetrically unique angle coordinates or, for TS, angles for which one of the subtending bonds is undergoing major bonding change, since the effects of bonding changes upon angle bending force constants is not well known. One caveat: the same coordinates should be present in the reactant and TS and this may dictate the choice of which angle(s) to delete. Equivalent results should be obtained from vibrational analyses employing different sets of coordinates if physically realistic force fields are employed. The literature from which force constants are taken must be consulted to determine which set of coordinates were used to derive the force constants in order to apply them properly.

If a complete set of $m(m-1)/2$ angle bending coordinates about each atom of bond multiplicity m are employed, then a Simple Valence Force Field (SVFF, a diagonal potential energy matrix – see Section III.B) generally provides a satisfactory approximation for KIE calculations[7,34]; an SVFF is desirable because it simplifies the interpretation of KIE. Removal of redundant coordinates alters the diagonal force constants and introduces significant off-diagonal force constants, so that use of a SVFF with a coordinate set from which redundant coordinates are simply deleted may be suspect. Extensive tests in our laboratory on a variety of molecular systems of the effect of simply deleting non-essential coordinates and employing a SVFF indicate that this approximation generally introduces no significant errors in KIE calculations. However, the procedure does introduce additional approximations and can be avoided by simply including all coordinates. The difficulty with this procedure is that the coordinate set for the TS will sometimes contain coordinates which are not present in either the reactants or products, which complicates the interpretation of KIE.

A modified procedure which we recommend is to include all coordinates in the reactant and product models, and to delete only those coordinates in the TS model which are not present in either the reactants or products (unless the coordinate is one of the linearly-dependent or symmetrically required coordinates of the TS model, in which case it must be retained). This procedure requires some modification of the formulation of the reaction coordinate motion and frequency, since eq. (35) was derived assuming that there were no redundant coorinates. If there are redundant coordinates, larger interaction force constants than implied by eq. (35) are generally necessary to produce the desired reaction coordinate frequency and motion. Initial estimates of interaction force constants may be obtained from eq. (35), and these increased in magnitude progressively until reaction coordinate frequency and motions are acceptable. The procedure of Yankwich and Buddenbaum[22b] provides a more systematic procedure for formulating reaction coordinate motion corresponding to a preselected frequency; incorporation of this procedure into the Bond Order method and into program BEBOVIB[39c] is in progress.

The unsolvated 8-atom TS (Fig. 6b) differs from the reactant model only in the bonding about *C; in this case $m = 5$ and $m(m-1)/2 = 10$ total angles, only $2m-3 = 7$ of which are linearly-independent. However, only one angle of the TS model, Y-*C-Cl, is not present in either the reactant (Ar*CH$_2$Cl) or the product (Ar*CH$_2$Y), and only this angle was excluded from the coordinate set of the TS. The coordinates of the TS thus include 7 stretching coordinates, one for each bond in Fig. 6b; 9 angle bending coordinates about *C; 3 angle bending coordinates about Ar_1; an out-of-plane bending coordinate for the groups bonded to Ar_1; and a torsional coordinate about the *C-Ar_1 bond - 21 coordinates in all, $3N-6 = 18$ of which are linearly-independent.

According to the Decius rules, inclusion of three solvating H-O groups in the solvated TS model requires addition of three Cl--H and three H-O stretching coordinates; three *C--Cl--H angles, three H--Cl--H angles; three pairs of Cl--H-O linear bending coordinates; and a *C--Cl torsional coordinate - in all 19 additional coordinates for a total of 40 vibrational coordinates for the solvated transition state model.

5. Force Constants for the Benzyl Chloride Models

Average standard values[38] of valence force constants for the benzyl chloride system are listed in Table 7. Stretching force constants for bonds of order other than unity were adjusted by the Pauling-Badger relation, eq. (17), using values of $F^o = F_R(1)$ from Table 7 and the assigned or assumed bond orders. Bending force constants were adjusted according to eq. (19) with x = 1/2, the geometry factor g(∅) of eq. (20), values of $F^o = F_{\emptyset}(1,1,T_d)$ from Table 7, and the assigned or assumed bond orders. Out-of-plane and torsional force constants were assumed not to change from their reactant values (listed in Table 7) in going to the TS.

TABLE 7. Valence Force Constants, F^o, employed for Benzyl Chloride Substitutions[a]

Bond	F^o(stretch)	Angle	F^o(Bend)
C-H	5.0	HCH	0.55
C-Cl	3.2	HCC	0.65
C-C	4.5	CCC	1.0
C-O	5.3	HCCl	0.60
Cl-H	5.2	CCCl	1.0
H-O	5.0	HCO	0.75
		CCO	1.25
		Torsion	
		C-C	0.072
		Cl-H	0.072
		Out-of-plane	
		C-Ar$_1$	0.20

[a]Units: md/A for stretching; mdA/rad^2 for bending, torsion and out-of-plane force constants. Values asssume single bonds and tetrahedral angles.

6. **Reaction Coordinate Motion for the Benzyl Chloride Reactions**

A zero or imaginary reaction coordinate frequency for the TS is achieved by introducing one or more interaction force constants, as in eq. (35). For the general S_N mechanism where either or both Y and the Ar group may contribute to displacement of Cl, asymmetric coupled motions both of C-Cl and Y-C (interaction coefficient a_{ij} of eq.(33) is referred to as A for this motion); and of C-Cl and Ar$_1$-C (interaction coefficient B) are included. If A (or B) is <u>positive,</u> r(Y-C) (or r(Ar$_1$-C)) decreases as r(C-Cl) increases as desired for reaction coordinate motion leading to products. The criterion for a zero (= sign) or imaginary (< sign) reaction coordinate frequency is that $(1-A^2-B^2)$ $\leq$ 0 (see eq. 34). The relative importance of intermolecular vs. intramolecular displacement is afforded by the relative values of A and B: A = 0 corresponds to displacement of Cl by Ar, as may be appropriate for an S_N1 reaction; B = 0 corresponds to displacement by nucleophile Y as in a concerted S_N2 process with no aryl participation.

7. Calculated KIE for the Benzyl Chloride Substitutions

The independent parameters of the model are n(*C-Cl), n(*C-Y), F_{Ar}, F_S, and the reaction coordinate parameters A and B. KIE for labeling with ^{14}C at *C, with deuterium at the benzyl (i.e., alpha) hydrogens, and with ^{37}Cl at the LG were calculated[100] for the independent parameters over the ranges of interest:

$0<$ {n(*C-Cl) or n(*C-Y)} < 1 and $0 <$ {n(*C-Cl) + n(*C-Y)} < 1

$0 < F_{Ar} <1$; F_S = 0.05 or 0.10 ; $0 <$ {$A^2/(A^2 + B^2)$} < 1

and for several different nucleophiles Y, and for temperatures between 25 and 60°C.

Figures 7, 8 and 9 summarize the carbon-14, deuterium and chlorine-37 KIE results, respectively, for Y = an oxygen atom cutoff model of the nucleophile, and for t = 25°C. The ordinate in Figs. 7 - 9 is the bond order n(*C-Cl), from 1.0 at the bottom to 0.0 at the top of each figure; the abscissa is n(*C-Y), from 0.0 at the left to 1.0 at the right.

In Fig. 7, contours of constant *C-14 KIE ($^{12}k/^{14}k$) are plotted as functions of n(*C-Cl), the vertical axis, and n(*C-Y), the horizontal axis. In all cases, F_S = 0.05, corresponding to solvation energy of ~6 kcal/mole per solvating group. The various plots in Fig. 7 correspond to different degrees of delocalization of the charge at *C into the ring (i.e., variation of F_{Ar}) and different reaction coordinate formulations (i.e., different values of A and B).

Fig. 7a corresponds to F_{Ar} = 0, no delocalization of charge at *C into the ring, and to B = 0, displacement by nucleophile only, with no contribution of displacement by the aryl group in the reaction coordinate. A Westheimer-Melander type bell-shaped behavior is evident along the central diagonal connecting reactants R and products P, and also along the diagonals parallel to this. These diagonals correspond to lines of constant tight-loose character t_x for the TS, so that the trend in KIE along any such diagonal from lower left toward upper right corresponds to the dependence of the KIE upon increasing product-like character P_x of the TS, analogous to Fig. 5 for hydrogen transfer reactions (see Sect. V). That S_N2 reactions can be regarded as "CH_2R" transfers between leaving group and nucleophile was first suggested by Fry[6] and by our earlier calculations for this system[43] and by calculations for S_N2 reactions of methyl halides by Willi.[105] Experimental evidence supporting this behavior has been reported by Ando et al.[106]

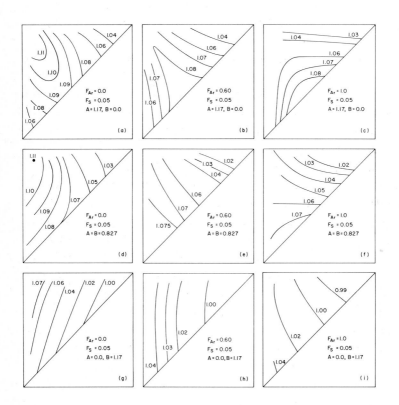

FIGURE 7. Contours of constant ^{14}C-KIE for S_N2 hydrolysis Reactions of Benzyl Chlorides.
     The ordinate is n(*C-Cl) and the abscissa is n(*C-O); reactant and product are located at lower left and upper right corners, and carbocation and adduct intermediates at the upper left and lower right corners, respectively, in each figure.

For a constant value of P_x, the isotope effect increases as t_x varies from 0.0 (central diagonal) toward $t_x = -1$ (upper left corner), corresponding to TS with increasing carbocation character; this trend is most easily seen in Fig. 7a along the secondary diagonal connecting carbocation and adduct.

Proceeding from left to right along the top row (Figs. 7a - c) corresponds to constant reaction coordinate motion (B = 0 in all cases, corresponding to displacement by nucleophile only) and an increasing fraction of charge delocalized into the ring, from 0% (F_{Ar} = 0, Fig. 7a) to 60% (F_{Ar} = 0.60, Fig. 7b) to 100% (F_{Ar} = 1.0, Fig. 7c). Since there is no charge developed at *C for processes with TS lying along the central diagonal, the intercepts of the contours with the central diagonal are the same in Fig. 7a - c. For TS with some degree of S_N1 character, i.e., those lying off the central diagonal, some charge is developed at *C and it is clear that conjugation with the ring, which represents additional bonding to *C, significantly reduces the *C-14 KIE. Proceeding from left to right in rows 2 and 3 corresponds to the same change in F_{Ar} as in row 1, and the same qualitative trend of reduction in *C-14 as F_{Ar} increases is evident (each row corresponds to a different, but constant reaction coordinate motion).

The left column in Fig. 7 (i.e., Figs. 7a, d, g) all correspond to F_{Ar} = 0. Fig. 7a corresponds to an S_N2-like reaction coordinate with displacement by nucleophile only (B = 0); Fig. 7d corresponds to equal contributions of displacement by nucleophile and by the aryl group (A = B); Fig. 7g corresponds to an S_N1-like reaction coordinate with displacement byaryl group only (A = 0). The same reaction coordinate parameters (A and B) apply to the corresponding figures in the center and right-hand columns. For any TS within Fig. 7a there is a rather substantial reduction of *C-14 KIE as the reaction coordinate motion (i.e., the displacement) changes from S_N2-like (Fig. 7a) to a mixed S_N2-S_N1 displacement (Fig. 7d) to an S_N1-like motion (Fig. 7g).

Very low values of *C-14 KIE, as observed for reactions believed to proceed by an S_N1 mechanism (see Table 4), are calculated for models in which displacement is by the aryl group (bottom row, Fig. 7), and particularly for TS structures with the greatest degree of delocalization of charge into the aryl ring (right-hand column, Fig. 7). Models for which the reaction coordinate motion corresponds to displacement by nucleophile (top row, Fig. 7) produce significant *C-14 KIE within the range observed for documented S_N2 processes. The model thus confirms the qualitative trends expected for S_N1 and S_N2 processes and produces calculated *C-14 KIE within experimental ranges.

Figure 8 displays calculated alpha-D_2 KIE for similar choices of parameters as in Fig. 7, except that the middle row (mixed S_N2-S_N1 reaction coordinate motion) and the middle column (60% delocalization, $F_{Ar} = 0.60$) are missing.

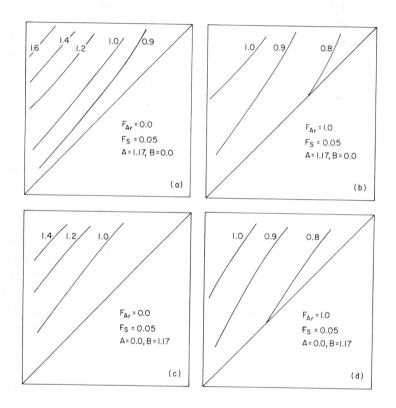

FIGURE 8. Contours of Constants alpha-D_2 KIE for S_N Reactions. Axes have the same meaning as in Figure 7.

The contours of constant KIE are nearly parallel to the central diagonal in all cases in Fig. 8, which indicates that the D-KIE is sensitive to t_x, the tight-loose character, but not to P_x, the reactant-product character of the TS. Indeed, plots of $k(H_2)/k(D_2)$ vs. t_x and constant P_x are very nearly linear with slopes that very slowly decrease as P_x increases. As for *C-14

KIE, smaller isotope effects are calculated for models with
substantial delocalization of charge (right-hand column, Fig. 8)
and for S_N1-like reaction coordinate motion (bottom row, Fig. 8),
although the decrease in KIE as F_{Ar} and B/A is increased is less
marked than for the corresponding trends in *C-14 KIE. The trends
in D-KIE are consistent with the suggestion[70] that alpha-D KIE are
determined predominantly by the net change in bending force
constants involving H/D atoms in going from R to TS. The
important determining parameters are thus the bending force
constants in the reactant (for angles H-*C-Cl, H-*C-H and
H-*C-Ar$_1$) and in the TS (for angles H-*C--Cl, H-*C-H, H-*C$\underline{--}$Ar$_1$
and H-*C--O). Since the bending force constants involving the LG
(H-*C-Cl angle bends) are smaller than those involving the Nu
(H-*C-O angle bends), the contours reflect slightly smaller KIE as
the TS becomes more product-like (i.e., as P_x increases).

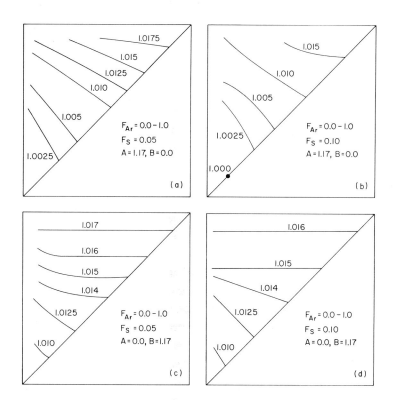

FIGURE 9. Contours of Constant ^{37}Cl-KIE for S_N Reactions.
Axes are the same as in Figs. 7 and 8.

Figure 9 displays contours of ^{37}Cl-KIE for the same parameters as for Figs. 7 and 8. Note here (see legends on Fig. 9) that the ^{37}Cl-KIE is independent of F_{Ar}.

For S_N2-like reaction coordinates (top row, Fig. 9), the ^{37}Cl-KIE are very nearly independent of t_x, the tight-loose character of the TS, but increase monotonically, but not linearly (the actual dependence is slightly sigmoidal), with P_x, and are very small for reactant-like and sizeable for product-like TS.

There is a rather dramatic reduction of the KIE for product-like TS as the solvation of the leaving group increases (left to right), but only a small reduction for reactant-like TS.

The ^{37}Cl-KIE are generally larger for an S_N1-like than for an S_N2-like reaction coordinate motion for a given TS structure. The calculated ^{37}Cl-KIE are nearly independent of n(*C-Nu) but dependent upon both n(*C-Cl) and n(*C-solvent) as expected (see Table 4 and earlier discussion).

Thus, the Bond Order model proposed for S_N reactions of benzyl chlorides agrees qualitatively with experimental trends, and calculated KIE are within the range of experimental results for the system. A more detailed analysis and comparison with experimental results for the benzyl chloride system is not possible because there are insufficient data for systems of constant composition and experimental conditions.

Similar Bond Order model KIE calculations have been carried out[107] on a series of bromide ion-alkyl bromide exchange reactions (Br$^-$ + R*CH$_2$Br for R = Me, Et, i-Pr, n-Pr, i-Bu, t-Bu, neo-Pentyl), yielding results in similarly good agreement with experiment as for the benzyl chloride system. Similar calculations[108] for the S_N2 reactions of benzyl arenesulfonates with N,N-dimethyl-p-toluidine yielded similar agreement with experiment, providing yet another example of the general reliability of the method.

The KIE calculated by the above Bond Order model for substitution reactions of benzyl chloride may be compared with those calculated from other models. Bron[109] has calculated *C-13 and alpha-D$_2$ KIE for S_N2 displacements of benzyl bromides. The structural model was similar to that above, except that a trigonal bipyramidal geometry was assumed. Bond lengths were adjusted by Pauling's rule (eq. 9), stretching force constants by the Pauling-Badger rule (eq. 17) and bending force constants by Johnston's rule (eq. 18). These features of Bron's calculations are all sufficiently similar to the corresponding features of the above Bond Order model to allow comparison of results. However, Bron employed a very different reaction coordinate formulation, resulting in larger imaginary frequencies for reactant-like and product-like TS than for symmetric TS; this is opposite to the trend suggested by early calculations by Johnston et al[110] and by the Hammond postulate.[59]

Buddenbaum and Shiner[80] have demonstrated in calculations of KIE for S_N2 reactions of methyl halides that the reaction coordinate motion for models employing Bron's formulation correspond to symmetric rather than asymmetric motion for some combinations of parameters. S_N2 displacements produced by a reaction coordinate using the Johnston[110] prescription, as employed in the above Bond Order calculations for benzyl chlorides, corresponded in the methyl halide calculations to a "rigid" CH_3 transfer with no H-*C-LG or H-*C-Nu angle bending (i.e., inversion about *C) accompanying the substitution. An appropriate reaction coordinate corresponding to concurrent *C-LG bond rupture, *C-Nu bond formation, H-*C-LG angle decrease and H-*C-Nu angle increase was formulated[80] using the reaction coordinate formulation of Buddenbaum and Yankwich.[79] Buddenbaum and Shiner emphasize that reaction coordinate freqency as well as reaction coordinate motion must be examined to assure that the TS model is physically reasonable.

In the Bond Order calculation of KIE for the benzyl chloride displacements, the reaction coordinate motion resulting from coupling only stretching of the *C-LG and *C-Nu bonds according to eq. (35) did involve $*CH_2R$ distortion in the direction of the expected inversion along with asymmetric displacement of LG by Nu.[100] The difference from the methyl halide case[80] is probably due to the difference in mass of H and R in the CH_2R and CH_3 groups. Although eq. (35) did produce a qualitatively realistic reaction coordinate motion for the benzyl chloride example, the method does not provide any simple or systematic way to vary reaction coordinate motion for a given of reaction coordinate frequency. The Buddenbaum, Yankwich formulation[79] does provide such a prescription, but applications to systems larger than the methyl halide substitutions are not simple. Incorporation of the Buddenbaum-Yankwich reaction coordinate treatment into the Bond Order method for modeling reactions would allow a systematic investigation of the contributions of reaction coordinate motion and frequency to kinetic isotope effects.

(C) S_N1 Model: The Tertiary-Butyl Chloride System

1. Mechanistic Summary

Both S_N1 and E1 mechanisms proceed by a rate-determining-step (rds) consisting of heterolytic rupture of the R*CH$_2$-LG bond to form an intermediate carbocation (presumed in most instances to be planar), which subsequently reacts with either a nucleophilic species Y to form an S_N1 product R*CH$_2$-Y, or with a base B to form the E1 products BH + olefin. The factors influencing the rate and consequences of the reactions, including the role of solvent and the effects of nucleophile (or base), leaving group, alkyl group, and stability of the carbocation intermediate have all been studied extensively and were discussed in the last section. In summary, it has been found that S_N1 and E1 processes are facilitated by a strongly solvating medium and by substrates which allow stabilization of the carbocation intermediate by extensive hyperconjugation or by conjugation with a π-bond system adjacent to the reaction center (i.e., tert-alkyl, allyl, or benzyl substrates). Kinetic isotope effect evidence for conjugative stabilization of the developing carbocation center is afforded, for example, by the observation of an inverse *C-13 KIE in the methanolysis of 1-bromo-1-(4'-methylphenyl)-ethane[111a] and an inverse *C-13 EIE for the ionization of triphenylmethyl chloride.[111b] Thus, provision for changes in hybridization, the role of solvent, and stabilization of the carbocation-like TS are all important features to include in modeling S_N1 processes.

The prototypical example of both the S_N1 and E1 processes is afforded by the solvolysis of tert-butyl chloride, eq. (47), and

$$(CH_3)_3{}^*CCl \xrightarrow{\text{rds}} (CH_3)_3{}^*C^+ + Cl^- \tag{47a}$$
$$(CH_3)_3{}^*C^+ + S^- \longrightarrow (CH_3)_3{}^*CS \quad (S_N1, \; S^- \text{ from solvent HS}) \tag{47b}$$
$$\longrightarrow (CH_3)_2{}^*C=CH_2 + HS \quad (E1) \tag{47c}$$

it is not surprising, therefore, that this reaction has been the subject of extensive study, including both the measurement and calculation of KIE. Experimental KIE for the substitution reaction have been determined for labeling at the alpha-carbon

(*C-14)[112], methyl hydrogens (D_1, D_2, D_3, D_6, D_9)[113] and chlorine leaving group (^{37}Cl),[114] and under conditions sufficiently comparable to afford a good example of the successive labeling approach[6]. The temperature dependence of the ^{37}Cl-KIE over the temperature range 10° - $60^{\circ}C$ has also been measured.[114]

The kinetics, particularly solvent effects on reaction rate, have provided an estimate, based on a simple electrostatic model, of a net charge development at *C in the TS of $+0.8e$, implying a late product-like TS (i.e., carbocation-like) with ~80% *C-Cl bond rupture. The observed ^{37}Cl-KIE ($^{35}k/^{37}k$ < 1.01) are much smaller than anticipated for this extent of bond rupture (values of ~1.02 are not unreasonable) - one of several examples in which LG-KIE are significantly smaller than expected on the basis of the extents of *C-LG bond rupture derived from other data.[99] There are two possible explanations for this discrepancy: either the extent of *C-LG bond rupture is smaller than indicated by kinetic data, or the bond loss occasioned by rupture of the *C-LG bond is partially compensated for by another source of bonding to Cl; solvent assistance of the LG has been suggested.[99] The *C-14 KIE is also small (~2.7%), consistent either with a small extent of *C-Cl bond rupture, or with a larger extent of bond rupture and stabilization of the charge by hyperconjugative interaction with the methyl groups. Such a stabilization is suggested by the significant methyl-deuterium KIE, averaging ~10% per deuterium.

2. Bond Order Models for KIE in tert-Butyl Chloride Solvolysis

Calculated *C-14, D_9 and ^{37}Cl-KIE have been reported[34] for a model similar to that described in the previous section for the S_N2 substitution reactions of benzyl chloride. The essential features of the reactant model are: standard bond orders of unity and single bond lengths (CH 1.09, CC 1.54, CCl 1.77 A) for all reactant bonds; tetrahedral angles (109.5°) about *C and each methyl-C; and the expected staggered configuration about the C-C bonds, with one anti and two gauche hydrogens, relative to the C-Cl bond, for each methyl group. This 14-atom model requires 40 vibrational coordinates (one *C-Cl, three *C-C_{Me} and nine C_{Me}-H stretches; three Cl-*C-C_{Me} and three C_{Me}-*C-C_{Me} angles bends about *C; three *C-C_{Me}-H and three H-C_{Me}-H angle bends about each C_{Me}; and a torsion about each C_{Me}-*C bond). Standard valence force constants (Table 7) were assumed for each coordinate.

The same coordinates are applicable to the simplest,

unsolvated TS model. The only formal bonding change in the rds, eq. (46a), is the rupture of the *C-Cl bond, the extent of which determines the charge build-up at *C, $+q = \{1 - n(*C--Cl)\}$. Some fraction P of this charge may be delocalized by hyperconjugative interaction with the methyl groups. This interaction can be thought of as donation of electron density from the C_{Me}-H bonds into the developing p-orbital at *C; the effect of this is to decrease the C_{Me}-H bond order and to effectively increase the *C-C_{Me} bond order. It is reasonable to assume that there is no net change in electron density about C_{Me} (see eq.(41) and related discussion preceding eq. (41)). Thus, ignoring for now any possible difference in hyperconjugative ability between gauche and anti hydrogens, the bond orders of the TS model become:

Bond	Bond Order
*C-Cl	$n(*C--Cl)$
*C-C_{Me}	$1 + \{1 - n(*C--Cl)\} \cdot P/3$
C_{Me}-H	$1 - \{1 - n(*C--Cl)\} \cdot P/9$

Bond lengths were adjusted by Pauling's rule (eq. 9) using the reactant values of $r(1)$ above and the bond orders derived from assumed values of $n(*C--Cl)$ and P, the independent parameters of the model. Bond angles about C_{Me} were assumed to remain tetrahedral and those about *C to vary linearly with the extent of bond rupture, as in eq. (48). Eq. (48b) is simply the geometric

$$\emptyset(Cl--*C-C_{Me}) = 90 + 19.5 \cdot n(*C--Cl) \tag{48a}$$
$$\cos\emptyset(C_{Me}-*C-C_{Me}) = 1 - 1.5 \cdot \sin^2\emptyset(Cl--*C-C_{Me}) \tag{48b}$$

relation between YMX and XMX angles for any YMX_3 structure of C_{3v} symmetry.

Stretching force constants were adjusted according to eq. (17) and bending force constants according to eq. (19) with $x = 1/2$, employing reactant values from Table 7 in both cases as the standard values. Torsional force constants were assumed to be unchanged from their reactant values. The *C-Cl stretching force constant was set to zero (flat barrier) or a small negative value (curved barrier) to produce a simple S_N1-type reaction coordinate corresponding to *C-Cl bond heterolysis.

Initial calculations indicated that values of $n(*C--Cl)$ and P (~0.2 and 0.6, respectively) which reproduced experimental *C-14 and D_9 KIE led to ^{37}Cl-KIE much larger than those observed. Conversely, values of n(*C--Cl) small enough to reproduce ^{37}Cl-KIE yielded *C-14 and D_9 KIE which were unacceptably low for any value of $0 \leq P \leq 1$. Satisfactory agreement with all KIE, including the temperature dependence of the ^{37}Cl-KIE, was obtained with a solvated TS model employing the same modeling as for benzyl chloride: three H-O solvating moieties, each bonded to Cl by a linear Cl--H-O hydrogen bond and occupying tetrahedral positions relative to the *C--Cl bond and staggered with respect to the three *C-C$_{Me}$ bonds; with n(Cl--H) ~ 0.05, corresponding to 6 kcal/mole solvation energy per H-O group.[34] The final values of n(*C--Cl) and P were 0.2 and 0.67, respectively, implying that $\emptyset(Cl--*C-C_{Me}) = 92^{\circ}$. An alternative model providing equally good agreement with experimental *C-14, D_9 and ^{37}Cl-KIE and its temperature dependence was provided by a Doering-Zeiss model[51] of weak specific backside nucleophilic assistance.[34]

Williams and Taylor[115] have reported calculations of the ^{37}Cl-KIE as a function of temperature using a 5-atom model (C$_3$*CCl) for both the reactant and non-solvated TS. A symmetrized force field fit to experimental reactant frequencies was employed, and the TS force constants for the symmetry coordinate corresponding most closely to *C-Cl stretching was set equal to zero to produce a reaction coordinate. The degree of planarity (angle C-*C-Cl) was an adjustable parameter, as were the other TS force constant values which, however, were not varied according to valence angle. Good agreement with experimental ^{37}Cl-KIE was obtained for r(*C--Cl) = 1.89 and r(*C-C) =` 1.50 A, corresponding to Pauling bond orders of 0.65 and 1.12, respectively, and $\emptyset(C*CCl) = 96^{\circ}$. In contrast to the above model, this is an early, reactant-like TS with much less *C--Cl bond rupture. The authors used the calculated KIE as evidence against solvation of the LG as being an important factor in the reaction; more recently, Taylor has supported solvent assistance in this reaction on the basis of other solvent studies.[116]

Although the Williams-Taylor model is improper for calculating *C-14 KIE, it seems certain that the bond orders, which imply P = 1.0 and hence complete compensation of charge at *C by increased *C-C bonding, would produce a negligibly small, perhaps even an inverse *C-14 KIE, compared to the experimental value of 1.027. The calculations of Taylor for this system reinforce the importance of investigating several kinetic isotope effects in successive labeling experiments to derive TS structures.

In addition to the D_9 KIE, Shiner et al[113a] measured the D_1, D_2, D_3, and D_6 KIE for the solvolysis of tert-butyl chloride and found that the KIE are non-cumulative (that is, $KIE(D_1) \neq \{KIE(D_n)\}^{1/n}$), presumably because of the different abilities of anti and gauche hydrogens to hyperconjugate. The experimental results allowed an estimate of the partitioning of the KIE into anti and gauche components $\{(^Hk/^Dk)_{anti} \sim 1.302; (^Hk/^Dk)_{gauche} \sim 1.011\}$. It is of considerable interest to determine whether the detailed analysis provided by calculations of KIE supports this interpretation.

The Bond Order model discussed above assumed no difference in force constants for H_a and H_g - i.e., no difference in abilities of anti and gauche hydrogens to hyperconjugate - and hence the calculated KIE_a and KIE_g differed only slightly due to the different geometries of H_a and H_g. Hyperconjugation should depend rather sensitively on the angle between the C-H bond and the p-orbital ($60°$ for gauche and $0°$ for anti hydrogens), favoring the ability of the anti hydrogen over that of the gauche hydrogens to hyperconjugate. If hyperconjugation of the gauche hydrogens are ignored relative to that of the anti hydrogen, the bond orders of the C-H bonds will be modified as below. With this minor

Bond	Bond Order
C-H$_{anti}$	$1.0 - \{1 - n(*C\text{--}Cl)\} \cdot P/3$
C-H$_{gauche}$	1.0

variation to the Bond Order model, the KIE are all brought into reasonably close agreement with the experimental values, as indicated in Table 8. The agreement between calculated values of D_a and D_g and those estimated from experimental KIE[113a] are probably within the limits of the experimental data to define the partitioning of the observed KIE into anti and gauche contributions.

3. Theoretical Calculations of tert-Butyl Chloride Ionization

The tert-butyl chloride substrate is simple enough to attempt a quantum mechanical investigation of the structural and bonding changes which occur upon heterolysis of the *C-Cl bond. Calculations in the absence of solvent correspond to gas-phase processes, quite distinct from the experimental conditions for

this solvolysis reaction (see ref. 117 for examples of
substitution reactions studied in the gas-phase). It is,
nevertheless, of considerable interest to see whether any evidence
of a hyperconjugative interaction involving the anti and/or gauche

TABLE 8. Comparison of Calculated and Observed KIE for the
Solvolysis of _tert_-Butyl Chloride

Label	Kinetic Isotope Effect	
	Calculated	Observed
*C-14	1.032	1.027
^{37}Cl	1.0104 (TD 15.4)[a]	1.010 (13.4)
D_9	2.160	2.327
D_6	1.672	1.7095
D_3	1.293	1.3303
D_2	(1.182)[b]	1.2022
D_1	(1.085)[b]	1.0922
D_a	1.2323	(1.3016)[c]
D_g	1.0239	(1.0110)[c]

[a] TD = temperature dependence = $10^4\{KIE(10^0) - KIE(60^0)\}$
[b] Calculated from D_a, D_g by method in ref. 113a.
[c] Estimated from D_2, D_1 in ref. 113a.

hydrogens is observed in such studies. We have recently carried
out MNDO calculations of the process described.[118] In these
calculations, all bond lengths and angles were optimized in the
reactant (resulting bond lengths: *C-Cl 1.8375, *C-C 1.5449, C-H$_a$
1.095, C-H$_g$ 1.084 A; bond angles: C*CCl 106.8^0, H$_a$C*C 110.8^0,
H$_g$C*C 111.6^0; staggered conformation, as assumed in above Bond
Order model). The *C-Cl bond was then stretched to successively
longer values and the remaining geometrical parameters allowed to
relax. At each positon, the charges on each atom were also
calculated from the electon densities.

The reactant was quite polar, with q = -0.225e localized on
Cl$^-$ and q = +0.225e distributed on the _tert_-butyl moiety. The
bond distances, bond angles, and distribution of charge in the

tert-butyl group are of particular interest and are shown in Figure 10 as a function of the Pauling bond order (eq. 9) of the *C-Cl bond. There is a considerable decrease in *C-C bond length and small but significant increases in C-H bond lengths, with a greater increase in C-H$_a$ than in C-H$_g$, as expected from a hyperconjugation argument (Fig. 10a). Bond angles about the methyl carbon do not change very much (Fig. 10b), but the direction of change is as expected if hyperconjugation of H$_a$ is greater than that of H$_g$. The linear decrease in $\emptyset$(*C*CCl) with n(*C--Cl) assumed in the Bond Order model is obviously not observed in these calculations, but the dependence observed here, if confirmed by more detailed calculations involving solvent, does not represent a serious defect of the Bond Order model since the KIE results are not very sensitive to bond angles.

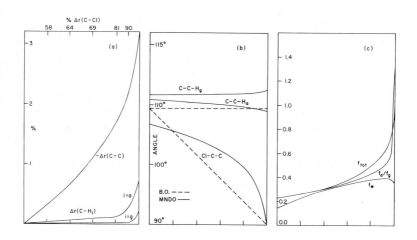

FIGURE 10. Parameters for the tert-Butyl Group from MNDO Calculations of the Gas-Phase Ionization of tert-Butyl Chloride

The charge distribution within the tert-butyl group (Fig. 10c) reveals that the fraction of charge localized on *C (f_*) increases as the *C-Cl bond is stretched (bottom curve), but less rapidly than the total charge separation (f_{tot}, top curve), which indicates an increasing tendency to delocalize the charge as the bond ruptures. Furthermore, the ratio of charge localized on the anti hydrogen to that on either gauche hydrogen (f_a/f_g) nearly coincides with the build-up of charge, f_{tot}; this suggests that the charge delocalized on the gauche hydrogens (which actually exceeds that on the anti hydrogen in the reactant) is a static or structural effect - perhaps a polarization effect of the negative Cl and the closer gauche hydrogens - whereas charge developed as the *C-Cl bond ruptures is preferentially accomodated on the anti hydrogen, as expected if the ability of the anti hydrogen to hyperconjugate exceeds that of the gauche hydrogens.

These calculations are not meant to confirm hyperconjugative stabilization in the solvolysis of tert-butyl chloride; the calculations are semi-empirical and certainly refer to a different (i.e., gas-phase) process. However, the results are consistent with hyperconjugative stabilization of the tert-butyl group as the *C-Cl bond ruptures in the absence of solvent. This system seems to represent a favorable case for a more detailed study employing ab initio theoretical methods and including solvent clusters to more realistically represent the process occurring in solution.[119] Impressive results have recently been reported using such methods for S_N2 reacions occurring in the gas-phase and in the presence of solvent clusters[120]. Such studies should provide a detailed examination of reaction processes, including kinetic isotope effects, for models of relatively simple processes occurring in solution. More importantly in the short term, such methods should provide information as to the reliability and applicability to TS of many of the empirical and semi-empirical relations used to mathematically model chemical reactions.

VII. APPLICATIONS TO ELIMINATION REACTIONS

Base-promoted 1,2-elimination reactions, eq. (49), form one

$$H'-\overset{H_2}{\underset{R_2}{C_2}}-\overset{H_1}{\underset{R_1}{C_1}}-X + B \longrightarrow \underset{R_2}{\overset{H_2}{>}}C_2=C_1\underset{R_1}{\overset{H_1}{<}} + BH' + X \qquad (49)$$

of the major classes of organic reactions, and are among the most extensively studied[16]. Mechanisms of these reactions are often classified[93,94] as irreversible E1 (rate-determining unimolecular heterolysis of C_1-X bond to form a carbocation intermediate, with subsequent rapid proton transfer to base); irreversible E1cb (rate-determining bimolecular proton abstraction by base, followed by unimolecular breakdown of the carbanion conjugate base of the substrate with rapid loss of X^-); and E2 (rate-limiting bimolecular proton abstraction by base with concurrent departure of the leaving group and double bond formation).

An E2 process in which the extents of bond rupture to the leaving group and to the transferring proton are equal is referred to as being "central" between the E1 and E1cb extremes. Such a process will be accompanied by double bond formation, which is often assumed to be, but need not be, equal to the extent of bond rupture. A spectrum of E2 processes[121] is envisaged, distinguished by the relative timing of C_2-H' and C_1-X bond ruptures, and bounded on one side by the irreversible E1 mechanism and extending through E1-like E2, central E2, E1cb-like E2, to the irreversible E1cb boundary. These various mechanisms are represented in Fig. 11 on a TS diagram for elimination reactions.[50,61]

Elimination reactions are especially well-suited to kinetic isotope effect studies,[61] particularly the successive-labeling approach.[6] At least four bonds suffer major bond order change during an elimination reaction (C_2-H', C_2-C_1, C_1-X and B-H'), and bond angle changes occur about both C_1 and C_2. However, the relative timing of bonding changes varies according to the mechanism, leading to different isotope effect consequences.[61] Thus, for an E1 process, significant KIE are expected for labeling at the positions about C_1 (1°- KIE for labeling at C_1 and X and 2°-KIE for labeling at H_1) but insignificant isotope effects are expected for labeling at any of the positions about C_2 (C_2, H_2, R_2 or H'). Conversely, in an E1cb process, significant isotope

effects for labeling at any position about C_2 (1^O-KIE for labeling
at C_2 or H'; 2^O-KIE for labeling at H_2) but no significant KIE
for labeling at positions about C_1 are expected. Significant KIE
could result from labeling at any position about C_1 or C_2 in an
E2 process, with magnitudes depending upon the relative E1 or
Elcb character of the E2-TS.

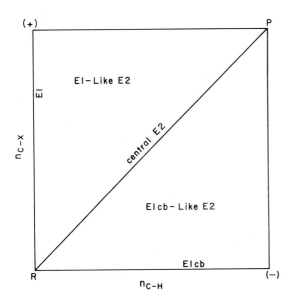

FIGURE 11. Transition State Diagram for Elimination Reactions

The observation of significant ^{14}C-KIE at both C_1 and C_2 positions
in an elimination reaction essentially rules out both the
irreversible E1 and Elcb mechanisms and provides strong supporting
evidence for an E2 mechanism. There are several recent examples
of this qualitative use of KIE in elimination reactions.[122] All
1,2-eliminations studied by the KIE method appear to proceed by E2
mechanisms; there are no examples where the carbon isotope effect
predictions for the irreversible E1 or Elcb mechanisms have been
realized.

The magnitudes of observed KIE for E2 processes, especially
when coupled with calculations of KIE,[7,122c] indicate that TS for
1,2-elimination processes have varying degrees of E1 or Elcb
character, depending upon the substrate, solvent/base system,
substituents at C_1 and C_2, and leaving group. Stereochemical and

other evidence indicates that both <u>anti</u>- and <u>syn</u>-eliminations may
occur, with <u>anti</u>-elimination being much more common.[16] Elimination reactions of phenylethyl derivatives are usually <u>anti</u>-
eliminations and proceed by E2 mechanisms with varying degrees of
Elcb-like character; this system will be used as an example of
modeling elimination processes in this section.

Proper cutoff models (Section IV) are employed for both
reactants and products, including a 3-atom cutoff model for the
phenyl group (<u>ipso</u> and the two <u>ortho</u> carbons), as in the S_N models
for the benzyl chloride solvolyses in Section VI. Both reactant
and product models are assumed to have standard tetrahedral angles
of 109.5° about sp^3 centers (including C_1 and C_2), trigonal planar
angles of 120° about sp^2 centers, and to exist in the most stable
staggered conformation about each single bond. Standard bond
orders of unity ($n=2$ for the $C=C$ bond in the olefin product) and
standard bond lengths[44] are assumed. Since there are four groups
bonded to both C_1 and C_2 by single bonds in the reactant, the
total bond order is 4.0 about each center. Coordinates are chosen
by the rules of Decius, as described in the previous section.
Standard valence force constants (Table 7 and ref. 38) are
employed. Figure 12 depicts the reactant and TS models for the
phenylethyl system.

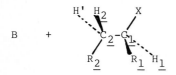

FIGURE 12a. Reactant Model   FIGURE 12b. TS Model
                                                for anti-elimination

R_1 = Ar, R_2 = H, 1-phenylethyl
R_1 = H, R_2 = Ar, 2-phenylethyl

For the TS model in Fig. 12b, there are 10 bonds (including
the 2-bond cutoff model of the aryl ring) to which bond orders are
to be asssigned. Other parameters of the model (bond lengths,
bond angles, force constants) are related to the bond orders and
corresponding parameters for the reactant and/or product models by
empirical or semi-empirical relations. Bond orders to the leaving
groups, $n(C_1-X)$ and $n(C_2-H')$ are obviously independent parameters

of the model, and must vary over the ranges of interest (~1.0 for reactant-like to ~0.0 for product-like TS in both cases) to encompass the mechanistic spectrum from E1-like to E1cb-like.

As for all hydrogen transfer processes (see Sect. V), the bond order about the transferring hydrogen H' is assumed to be unity during the course of the reaction, as in eq. (50):

$$n(B-H') = \{1 - n(C_2-H')\} \tag{50}$$

The C_2-H_2 bond to the non-transferring hydrogen(s) and the C_1-H_1 bond(s) undergo no significant bond order change during the course of the reaction. There is a rehybridization at C_1 and at C_2 which will be accompanied by changes in angle bending force constants, and there will be very slight decreases in bond length in the transformation from $C(sp^3)-H$ to $C(sp^2)-H$ bonds and a corresponding increase in stretching force constant. Since these changes are small, and since the KIE involving these hydrogens are expected to depend primarily upon changes in bending force constants (which are more sensitive to the changes in valence angles than to small changes in bond order - see Table 3, Fig. 3 and eqs. (19) and (20)), the small bond order order changes associated with rehybridization are generally ignored, as they were for S_N models in Sect. VI, and the reactant bond orders retained in the TS:

$$n(C_i-H) = 1.0 \qquad \text{for both R and TS} \tag{51}$$

A refinement in the model is to retain relevant bond orders at the reactant value of unity, but to adjust force constants (the important parameters in KIE calculations) according to the hybridization of the atoms involved. Bending force constants are adjusted for hybridization by eqns. (19) and (20). Equations (52a) and (52b) are empirical relations which provide reasonably good estimates of how single bond stretching force constants for C-H and C-C bonds change with hybridization of the carbon atoms in

$$F^{O}(C_i-H) = 4.6 - 1.2\cos\emptyset_i \qquad \text{mdA}^{-1} \tag{52a}$$
$$F^{O}(C_i-C_j) = 3.5 - 1.5(\cos\emptyset_i + \cos\emptyset_j) \qquad \text{mdA}^{-1} \tag{52b}$$

stable molecules. The angles $\emptyset_i$ and $\emptyset_j$ assume the standard values of $109.5°$ for sp^3, $120°$ for sp^2 and $180°$ for sp hybridization at

C_i and C_j, respectively. A model for varying $\emptyset_i$ and $\emptyset_j$ in the TS will be presented below (eqs. 58a,b). The values of F^o from eqs. (52a,b) are used as the standard values for reactants and products and in the Pauling-Badger relation (eq. 17) for the TS and for bonds of order other than unity.

The aryl group at C_i may serve to delocalize and stabilize the charge developed at C_i by conjugation. This involves an increase in the quinoid-type resonance structures which leads to an increase in the C_i-Ar_1 bond order and a decrease in the Ar_1-Ar_2 bond orders, such that $N(Ar_1)$, the total bond order at the Ar_1 position, remains 4.34, the value for an aromatic carbon center,[47] eq. (53); see also eq. (41) and associated discussion.

$$n(Ar_1-Ar_2) = \{N(Ar_1) - n(C_i-Ar_1)\}/2 \qquad (53)$$

In an Elcb-like E2 process, bonding loss to the transferring hydrogen $\{\Delta n(C_2-H') = 1 - n(C_2-H')\}$ runs ahead of bonding loss to the leaving group $\{\Delta n(C_1-X) = 1 - n(C_1-X)\}$, and the extent of double bond formation is limited by the negative charge density developed at C_1. For E1-like E2 processes, double bond formation is limited by the extent of proton transfer and development of positive charge density at C_2. If $f_{C=C}$ is the fraction of the bond loss at C_i which is compensated for by double bond formation in the TS, then the bond order of the C-C bond will be given by eq. (54). A value of $f_{C=C} < 1.0$ implies a "loose" TS in which

$$n(C_1-C_2) = 1 + f_{C=C} \cdot \Delta n(C_i-LG_i) \qquad (54)$$

where $i = 1$ and $LG_1 = X$ for an Elcb-like E2 process
          $i = 2$ and $LG_2 = H'$ for an E1-like E2 process

the extent of double bond formation does not completely compensate for bond loss to the LG (since the C_1-C_2 bond order is not included in Fig. 11, the "tight-loose" character cannot be represented on the elimination TS diagram). There may be several reasons why the bonding of the TS may be loose and $f_{C=C} < 1$. One obvious reason is that steric crowding in the TS may preclude proper alignment (dihedral angle $\tau = 0^o$) of the developing p-orbitals at C_1 and C_2. In the TS, the total bond order N_i about reacting centers C_i, $i = 1,2$, is the sum of the bond orders of the four groups (ligands, L) bonded to C_i, as in eq. (55). As

$$N_{\underline{i}} = \sum_{j=1}^{4} n(C_{\underline{i}}-L_{\underline{j}}) \leq 4.0 \tag{55}$$

a consequence of bonding changes in going from reactant ($N_{\underline{i}}$(Rcnt) = 4.0 for both i = 1 and 2), a partial charge $q_{\underline{i}}$ ($0 \leq q_{\underline{i}}/e \leq 1$) will develop at $C_{\underline{1}}$ and/or $C_{\underline{2}}$ (positive and negative, respectively) according to eq. (56). Some fraction f_{Ar} of this charge may be stabilized by interaction with the phenyl group at $C_{\underline{2}}$ (for

$$\left|q_{\underline{i}}\right| = 4.0 - N_{\underline{i}} \qquad i = 1,2 \tag{56}$$

the 2-phenylethyl case) or at $C_{\underline{1}}$ (for the 1-phenylethyl case). Thus, eq. (57) gives the bond order of the $C_{\underline{i}}$-aryl bond.

$$n(C_{\underline{i}}-Ar_{1}) = 1.0 + f_{Ar} \cdot \left|q_{\underline{i}}\right| \tag{57}$$

The independent bond order parameters of the model are thus $n(C_{\underline{1}}-X)$, $n(C_{\underline{2}}-H')$, $n(C_{\underline{1}}-C_{\underline{2}})$ or $f_{C=C}$, and $n(C_{\underline{2}}-Ar_{1})$ or f_{Ar}. The bond orders for the remaining bonds (B--H', $C_{\underline{i}}-H_{\underline{i}}$, $Ar_{1}-Ar_{2}$) are fixed by the independent bond orders or assumed mechanistic constraints. Specification of valence angles, reaction coordinate motion, and the possible inclusion of specific solvation are needed to complete the model.

As the elimination reaction proceeds, the hybridization of the orbitals used to bond the LG (H' at $C_{\underline{2}}$ and X at $C_{\underline{1}}$) changes from sp^{3} to p. During this transformation from reactants to products, the (idealized) angles $\emptyset_{\underline{i}}$ between the non-reacting groups bonded to $C_{\underline{i}}$ retain a three-fold symmetry with respect to the LG. It is reasonable to assume that a three-fold symmetry is maintained also in the TS, and to require the LG to depart at an angle $\theta_{\underline{i}}$ with respect to the C-C bond that preserves this symmetry. The requirements are different at $C_{\underline{1}}$ and at $C_{\underline{2}}$. At $C_{\underline{1}}$, a positive charge develops and $C_{\underline{1}}$ acquires some carbocation character. Since carbocations are expected to be planar trigonal, the angles $\emptyset_{\underline{1}}$ should be 109.5° in the reactant and 120° in the product; in the TS, the angles should be intermediate between these extremes, in proportion to the carbocation character developed at $C_{\underline{1}}$. At $C_{\underline{2}}$, on the other hand, carbanion character develops during the reaction. If the negative charge is delocalized by conjugation with the aryl group, the angles $\emptyset_2$ should approach $120°$ as for conjugated alkyl anions; if the charge is not delocalized, the groups about C_2 should remain tetrahedral as in the reactant. Equations (58a) and (58b) provide such behavior for the angles $\emptyset_1$ and $\emptyset_2$. The angles θ_i are calculated from $\emptyset_i$ using the geometric relation in eq. (58c) - see also eq. (48b) and the related discussion.

$$\cos\emptyset_i = \{-1.0 + f_i{}^x(n)/3\}/2$$

where
$$f_1 = n(C_1-X) \tag{58a}$$
$$f_2 = \{(2-n(C_1-C_2))\cdot(2-n(C_2-Ar_1))\cdot$$
$$(3-n(C_1-C_2)-n(C_2-Ar_1))\cdot n(C_2-H')\} \tag{58b}$$

$$\cos\theta_i = 1 - 1.5\cdot\sin^2\emptyset_i \tag{58c}$$

In eq. (58), f_i can be regarded as the fractional change in angle $\emptyset$ ($0 \le f_i \le 1$) as the product-like character of the TS increases. Fig. 13 shows the behavior of angles $\emptyset_i$ for values of x = 1, 1/2, 1/4. Values of x = 1/2 provide somewhat better agreement, when combined with eqs. (19) and (20) to adjust bending force constants, with deuterium isotope effects - see for example ref. 123.

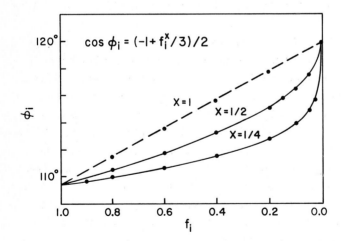

FIGURE 13. Variation of Angles $\emptyset$ with f_i of Eq. (58)

Reaction coordinate motion for an E2 reaction is usually considered to correspond primarily to alternating compression and extension of the B--H', H'--C_2, C_2-C_1, and C_1-X bonds (referred to as bonds 1, 2, 3, 4 below), respectively. There may, of course, be some contribution from C_i-Ar_1 bond compression if delocalization of charge at C_i occurs, and contributions from angle bending of the non-reacting groups at C_1 and C_2. Saunders[124] first reported calculations of KIE in elimination reactions using the alternating four-bond formulation for reaction coordinate motion, which provides an example of the general method outlined in Section IV.C. Interaction force constants f_{ij} = $a_{ij}(F_{ii}F_{jj})^{1/2}$, with a_{ij} > 0, were used to couple adjacent bond stretches to produce the alternating motion(see eq. 33). If A represents the interaction coefficient (a_{ij} above) for coupling B--H' and H'--C_2 stretches, B that for coupling H'-C_2 and C_2-C_1 stretches and C that for coupling C_2-C_1 and C_1-X stretches, eq. (35) yields for this case:

$$(1 - A^2 - B^2 - C^2 + A^2C^2) = D \leq 0 \qquad (59)$$

A^2/C^2 is a measure of the importance of hydrogen transfer to leaving group expulsion in the reaction coordinate. Saunders[124] found that calculated KIE for E1cb-like E2 reactions are (not surprisingly) within experimental range only for a reaction coordinate dominated largely by hydrogen transfer: A ~ 1.0, B = C ~ 0.3.

Specific solvation of the leaving group, the attacking base, and the charge centers in the substrate, C_1 and C_2, should all be taken into account in a realistic model. In fact, solvation at all of these centers may not be necessary to take into account, either because intimate solvation may be unlikely because of steric requirements, or because the KIE are not sensitive to the small differences in solvation energy between the reactant and transition state. Comparisions of KIE calculations for unsolvated models with experimental results will reveal deficiencies in the models, and chemical insight into the possible importance of solvation is used to revise the models to bring calculated and experimental KIE into better agreement. The specific example discussed below will provide an example of how solvation may be included.

All calculations of KIE for elimination reactions reported since Saunders' original calculations for elimination reactions[124] have employed similar models and reaction coordinate formulations (indeed, similar values of A, B, and C). These include KIE calculations for dehydrochlorination of 1,1-diaryl-2,2-dichloroethanes by alkoxide bases,[125] E2 eliminations of chloroalkanes by hydroxide and alkoxide bases,[126] base-promoted eliminations of substituted 1-phenylethyl chlorides,[122c,127] Hofmann eliminations from substituted (2-phenylethyl)trimethylammonium bromides[123,128], base-promoted elimination of HCl from 2-phenylethyl chlorides[129]. Good agreement with experiment was generally achieved in these calculations, lending support to the applicability of the Bond Order method to elimination reactions.

The calculations of KIE in the Hofmann eliminations from (2-phenylethyl)trimethylammonium ions[123,128] illustrates some additional features of Bond Order modeling of eliminations. Some of these features have been mentioned already: the reactant model employed standard tetrahedral angles about all aliphatic carbons and the nitrogen, and standard planar trigonal angles for the 3-atom cutoff model of the aryl group; standard bond lengths and bond orders of 1.0 for single bonds and 1.67 for aromatic C-C bonds. An all-staggered conformation with the aryl ring anti to the trimethylammonium leaving group was assumed. The standard set of valence coordinates (see Sect. VI) and standard valence force constants (Table 7 and ref. 38) were employed. The TS model involved an alkoxide base (C-O$^-$ cutoff model) attacking along the C_2-H' bond to form a linear C_2--H'--O hydrogen bond with conservation of bond order about H'; the H'--O-C_{base} angle was assumed to be tetrahedral with a single O-C_{base} bond. Bond orders of the B--H', H'--C_2, C_2-C_1 and C_1-N bond were the independent bond order parameters of the model.

In the TS for this documented anti-elimination, the aryl group must assume a gauche positon with respect to the bulky trimethylammonium leaving group, causing steric crowding which may preclude proper alignment (dihedral angle τ = 0°) of the developing p-orbitals at C_1 and C_2. For a very reactant-like TS, the dihedral angle may be close to 45°, the limiting value above which the syn-hydrogen becomes better disposed for elimination. For a product-like TS, perfect alignment (τ = 0°) should be approached. The dihedral angle, therefore, should be related to both the C_1-N and the C_2-H' bond orders. Eq. (60) provides a proper behavior for τ. Since the orbital overlap is proportion-

$$\cos\tau = \{0.5 + 0.5\,\Delta n(C_1-N)\,\Delta n(C_2-H')\}^{1/2} \tag{60}$$

al[130] to $\cos \tau$ and the π-bond order is proportional[131] to $\cos^2 \tau$, the fraction of the limiting bond rupture, $\Delta n(C_1-N)$, which is compensated for by double bond formation is, in this case, $f_{C=C} = \cos^2 \tau$ (see eq. 53). Other features of the TS are defined by eqs. (50) – (57).

The fraction of the charge developed at C_2 which is delocalized by the aryl group at that position, f_{Ar} in eq. (57), was treated as an independent parameter of the S_N model for reactions of benzyl halides. It is reasonable, however, to assume that the ability of the aryl group to stabilize charge should vary with the ability of the substituent Z carried by the ring to accept or supply electron density. This ability is often expressed by the Hammett parameter σ_Z or one of the other substituent parameters.[132] It is thus reasonable in these reactions to assume that $f_{Ar} = K_i \cdot \sigma_Z$, where the proportionality constant K_i depends upon the reaction system and the nature of the aryl group. For example, for the Elcb reactions of 2-phenylethyl derivatives, the negative charge developed at C_2 may be delocalized by the substituted phenyl ring bonded to C_2. No data are available relating to charge delocalization in $PhCHCH_2X^-$ anions, but NMR data[133] for metal benzyls provide an estimate of ~50% charge delocalization in the related parent benzyl anion ($PhCH_2^-$, where the substituent on the Ph ring is H, for which $\sigma_H = 0$). Thus, a realistic approximation for the 2-phenylethyl system is that $f_{Ar} = 0.5(1 + \sigma_Z)$, which leads to 50% delocalization for the parent system to ~90% delocalization in the case of a para-NO$_2$ substituent.

A four-atom reaction coordinate formulation, including Saunders'[124] values of A, B and C (see eq. (59) and related discussion) were employed. Initial results indicated acceptable agreement with D_2-KIE at position C_2 (both 1° and 2° KIE contribute to the overall result) and with ^{15}N KIE were obtained for reasonable values of the parameters of the TS model. However, the calculated D_2-KIE at position C_1 were somewhat too large, and the calculated methyl-deuterium KIE (D_3, D_6 and D_9 have all been measured[134] and average ~4% per CD$_3$ group) were less than 0.1 of those observed. The C_1-D_2-KIE conceivably could be brought into agreement by changes in bending force constants to those atoms, although there is no reason to believe that very different force constant values than those employed should be expected. There are few possibilities to account for the sizeable methyl-deuterium KIE observed, however. An inversion about the N atom as the LG departs could conceivably cause sufficient changes in force constants to result in observable effects; alternatively, desolvation of the leaving group (which carries a positive charge in the substrate) as it departs as a neutral species might also contribute to the observed effects. Consequently, at the exposed trimethylammonium and C_1 positions, solvation groups were added to explore the possible importance of solvation to observed KIE. Solvation of the nitrogen (which carries the charge in the substrate) or of C_2 was deemed to be unlikely because of steric crowding, and solvation of the base need not be considered since no KIE for labeling at positions sufficiently close to the base to reflect this bonding have been measured.

Solvation of the trimethylammonium moiety probably is a result of small C_{Me}-H dipoles induced by the proximty of the charged N atom. If this is so, specific solvation would reflect the local three-fold symmetry of the methyl groups, allowing modeling by a single solvating group along the axis of each methyl. Similarly, a single solvating group bisecting the C_1H_2 group was used to model specific solvation of the developing carbocation at C_1. A single oxygen-atom cutoff of the solvating groups was employed, with solvation bond orders corresponding to solvation energies of less than 10 kcal/mole in the reactant (see eqs. (43) and (44) and related discussions.

Good agreement with all available KIE (C_2-D_2, C_1-D_2, ^{15}N, and methyl-deuterium) for the (2-phenylethyl)trimethylammonium system was obtained for the solvated model.[123] The TS were found to be located in the Elcb-like E2 region on a TS map (see Fig. 11), along a line parallel to the central diagonal (principal diagonals in Fig. 11 correspond to TS with constant charge q^- developed at C_2, in this case q^- ~0.2e) with a nearly central product-like character for the parent compound and increasing (decreasing) product-like character for more electron-donating (electron-withdrawing) substituents.

[14]C-KIE were also calculated for labeling at both C_1 and C_2, although experimental data were not available at the time. Both KIE have since been determined[128] and reveal that revision of the model is required to produce agreement with the new results. We have recently obtained agreement by varying the reaction coordinate formulation to allow more heavy-atom motion, relative

to hydrogen transfer, as the para-substituent on the aryl ring becomes more electron-donating and the TS becomes more product-like. This is exactly what one would expect for reaction coordinates which proceed smoothly from reactant through the various TS to products, as for example, RC-C in Figure 4, Sect. IV.C (see also Fig. 8 in ref. 123). There is recent evidence that bending motions must be explicitly included in the reaction coordinate formulation for elimination reactions in order to account for the magnitudes and temperature dependencies of H/D and/or H/T KIE in 1,2-elimination reactions.[135]

Carbon-14 KIE for the syn-elimination of N,N-dimethyl-hydroxylamine from (2-phenylethyl)dimethylamine oxides have recently been reported.[136] The beta D_2-KIE has also been measured for the reaction.[137,138] This reaction involves either a direct intramolecular H-transfer or a transfer involving an intervening solvent molecule.[138] Preliminary calculations[139] involving a cyclic TS without intervening solvent and a reaction coordinate corresponding largely to hydrogen transfer, but including explic-itly H'CC and CCN angle bending to bring the transferring H' into closer proximity to the oxygen transfer terminus, indicate that satisfactory agreement with observed KIE can be obtained, so that the intervention of solvent does not appear to be an obligatory requirement of the TS model. Further calculations including solvent are in progress.[140]

VIII. APPLICATIONS TO REARRANGEMENT AND OTHER REACTIONS

Rearrangement reactions often involve cyclic transition states, even though the reactants (and usually products also) are generally acyclic. A cyclic TS presents a special difficulty for KIE calculations, since it is not possible to use the same coordinates for the vibrational analysis of the TS as for the reactant, and the changes in force constants are not easy to interpret. Keller and Yankwich[141] have provided a general treatment of redundant coordinates in cyclic TS, and they propose either i) eliminating some coordinates to remove the redundancies (see, however, discussion in Sect. VI relating to force fields) or ii) transforming the force field to correspond to a non-redundant $\underline{G}$ matrix[32]. The methods proposed for accomplishing ii) are not easy to apply and have not been incorporated into the Bond Order method or program BEBOVIB.[39c]

The Bond Order method has been applied to the rearrangement accompanying solvolysis of neophyl arenesulfonates,[2] for which a simple solution to the cyclic TS problem was adopted. If the actual reaction proceeds according to eq. (61), the problem of

$$\text{acyclic reactant} \longrightarrow \{\text{cyclic TS}\} \longrightarrow \text{acyclic products} \qquad (61)$$

cyclic redundancies in the TS occurs. If the reaction is considered to occur as in eq. (62), then the KIE calculated for the second step involving cyclic models for both reactants and TS

$$\text{acyclic reacant} \Longleftrightarrow \{\text{cyclic reactant}\} \longrightarrow \{\text{cyclic TS}\} \qquad (62)$$

will represent those for the overall process, eq. (61), provided that the isotope effects for the "pseudo equilibrium", represented by the first step of eq. (62), are unity for all labeled positions. This technique really consists of choosing for the reactant a vibrational coordinate system which is easier to relate to the TS coordinates than is the coordinate system generated by the Decius rules. As mentioned in Sect. VI, the same frequencies should result when different coordinate systems are employed for the same molecule, provided the force constants are related properly. This is too stringent a requirement for KIE calculations, and it is sufficient that KIE for the "pseudo equilibrium" involving structures defined by the two coordinate systems to be unity.

In the neophyl rearrangement calculations,[2] an aryl group migrates to an adjacent carbon as the leaving group departs, so that the aryl group forms a "bridge" between the adjacent carbons and bonds to both in the TS. The "pseudo equilibrium" assumed in this case is simply that relating the reactant $Ar-CR_2-CH_2-LG$ and the cyclic analogue CR_2(bridged Ar)CH_2-LG. This is equivalent to replacing the $AR-C_2-C_1$ angle bend with a new stretching coordinate between $Ar-C_1$, other coordinates remaining the same. It is possible to derive an analytical expression between these two coordinates; however, other coordinates of the model, and especially force constants other than those related to the two valence coordinates, are also afffected by this change of representation. The preferred procedure is to make the change of coordinates and corresponding force constants, and to systematically adjust other elements of the force field until KIE calculated for the "pseudo equilibrium" are unity at all positions of interest. Good agreement between experimental and calculated KIE was obtained in the neophyl rearrangement study for reasonable TS structures and involving the acyclic ⟷ cyclic "pseudo equilibrium"described for the reactant. Use of the Keller-Yankwich procedure[141] would allow a more systematic treatment of cyclic TS in such reactions.

Bowie, et al[86] have carried out Bond Order calculations of H/D KIE accompanying the McLafferty rearrangement observed in mass spectral studies, as well as calculations of H/D KIE in the elimination of propanol from $EtO^-/MeCO_2Pr$ in the gas-phase. The results are in good agreement with experiment and provide a good example of the application of the method to this type of reaction.

Recently, Kanska and Fry reported the first measured value of a carbon isotope effect in a simple electrophilic addition reaction between 2,4-dinitrobenzenesulfenyl chloride and both 1- and 2-^{14}C-labeled styrene.[122a] Addition reactions to double bonds are the reverse of elimination processes and the interpretation of KIE in such reactions should be closely related to those for elimination reactions. The mechanistic possibilities and the large number of positions at which bonding changes occur make these reactions attractive for study by the KIE technique. Calculations of the KIE for this addition reaction, assuming two distinct mechanisms (one involving an open-chain and one a cyclic TS) are in progress.

APPENDICES

(A) Development of Statistical Mechanical Expressions for IE

The energetics of a chemical reaction which converts reactants R_i into products P_j, proceeding through an activated complex or transition state TS, may be represented by a reaction diagram as in Figure 2b, in which the potential energy E is plotted along the least-energy path connecting reactants and products, the reaction coordinate. According to statistical thermodynamics, the equilibrium constant K and the rate constant k for the reaction may be related to the partition functions per unit volume, Q, of the various species by eqs.(Al) and (A2), where the

$$K = \exp(-\Delta G^{o}/RT) = (Q_P/Q_R) \cdot \exp(-\Delta E_o/RT) \qquad (Al)$$

$$k = \exp(-\Delta G^{TS}/RT) = (k_b T/h)(\tau \gamma)(Q_{TS}/Q_R) \cdot \exp(-\Delta E^{TS}/RT) \qquad (A2)$$

partition functions are to be evaluated relative to the potential energy extrema for R, P, and TS. The partition function Q_{TS} differs from that for stable molecules such as R and P in that the normal vibrational mode corresponding to motion along the reaction coordinate is unstable, and corresponds to internal translation, or separation of the products of the reaction from each other as the reaction proceeds; the corresponding reaction coordinate frequency ν_L will be zero or imaginary $\{(\nu_L)^2 \leq 0\}$, and the contribution of this mode is factored out of Q_{TS} and leads to the term $(k_b T/h)$ in eq. A2[19a]; k_b and h are the Boltzmann and Planck constant, respectively, and T is the Kelvin temperature. The terms τ and γ are the quantum mechanical tunneling and transmission coefficients, respectively.[19a]

According to the Born-Oppenheimer approximation,[87] the potential energy E does not change when one or more atoms in one of the reactants R is substituted by another isotope of that atom to form a labeled molecule *R (an isotopic isomer of R), but both the equilibrium constant K and rate constant k may change. Molecular dynamics calculations indicate that the transmission coefficient γ is generally isotope-insensitive,[87] and the isotope effect on the tunneling factor τ is discussed in section IV.C; therefore only the effect of isotopic substitution on the partition functions needs to be considered further here. For the reaction of the isotopic isomers R and *R, proceeding through transition states TS and *TS to products P and *P, application of eq. (A1) and (A2), assuming the molecules to be independent and all molecules of the same isotopic species to be indistinguishable (which is strictly valid only for ideal gases; see Section II for a discussion of these approximations), yields eqs. (A4) and (A5) for isotope effects upon the equilibrium constants (EIE) and upon the kinetic rate constant (KIE). The molecular partition

$$EIE = K/*K = (Q_P/Q_{*P})/(Q_R/Q_{*R}) \qquad (A4)$$
$$KIE = k/*k = (Q_{TS}/Q_{*TS})/(Q_R/Q_{*R}) \qquad (A5)$$

function Q is the sum of Boltzmann factors $\{g_i \exp(-e_i/k_bT)\}$ over all energy states i of energy e_i and degeneracy g_i for the species in question.

To evaluate the partition functions, the molecular energy is generally assumed to be separable into translation, rotation, vibration, and electronic contributions, so that the partition function is factored, as in eq. (A6) and (A7). The separate

$$e = e_{trans} + e_{rot} + e_{vib} + e_{elec} \qquad (A6)$$
$$Q = Q_{trans} \cdot Q_{rot} \cdot Q_{vib} \cdot Q_{elec} \qquad (A7)$$

factors are generally evaluated within the rigid rotor, harmonic oscillator approximation,[87] which yield the familiar expressions:

$$Q_{trans} = (2\pi m k_b T/h^2)^{3/2} \qquad (A8)$$

$$Q_{rot} = (\pi)^{\{(n_{rot}-2)/2\}} \cdot \prod_{i=1}^{n_{rot}} (8\pi^2 I_i k_b T/h^2)^{1/2} \qquad (A9)$$

$$Q_{vib} = \prod_{i=1}^{n_{vib}} \{\exp(-u_i/2)/(1 - \exp(-u_i))\} \qquad (A10)$$

$$Q_{elec} = g_0 = 1.0 \text{ for most molecules} \qquad (A11)$$

n_{rot} and n_{vib} are 2 and 3N-5 and 3 and 3N-6 for linear and non-linear N-atom molecules, respectively.

Substitution of eqns. (A7 - A11) into eqns. (A4) and (A5) yields the primary expressions, eq. (2a, b, c) and eq. (3) of section II.

Isotopic isomers obey the Teller-Redlich product rule[25] eq. (A12); see also appendix B. Realizing that the first factor in

$$\prod_{i=1}^{N}(m_i/{}^*m_i)^{3/2} \cdot \prod_{i=1}^{n_{vib}}(u_i/{}^*u_i) = (M/{}^*M)^{3/2} \cdot \prod_{i=1}^{n_{rot}}(I_i/{}^*I_i)^{1/2} \qquad (A12)$$

eq.(A12) will cancel in the ratio of any two isotopic species (since they contain the same atoms, by definition), and defining the vibrational frequency product ratio, VP as in eq. (A13),

$$VP = \prod_{i=1}^{n_{vib}}(u_i/{}^*u_i) \qquad (A13)$$

substitution of eq. (A12) and (A13) into eq. (2) of the text leads to eq. (4a, b) of Section II.

(B) <u>Heuristic Derivation of the Teller-Redlich Product Rule</u>

The motion of a collection of N particles may be described in any coordinate system in which the equations of motion may be formulated and solved. The solutions of the equations (the allowed energy levels) are properties of the molecule and not of the particular coordinate system used to set up the equations of motion.[32, 87] In any coordinate system the kinetic energy (T) and the potential energy (V) of a molecule of N atoms will be (to first approximation) quadratic functions of the coordinates q and conjugate momenta p ($p_i = \partial T/\partial \dot{q}_i$, where $\dot{q}_i = dq_i/dt$):

$$T = (1/2)\sum_{i,j} t_{ij}p_ip_j \qquad V = (1/2)\sum_{i,j} f_{ij}q_iq_j \qquad (B1)$$

or, in matrix form:

$$2T = \underline{p}^t\underline{T}\underline{p} \qquad 2V = \underline{q}^t\underline{F}\underline{q} \qquad (B2)$$

where $\underline{p}^t$ is the transpose of the column matrix $\underline{p}$ of momenta, etc.

The Lagrange equations of motion for the system, assuming simple harmonic motion, then yield the secular equation:

$$\underline{TFL} = \underline{L}\,\underline{\Lambda} \qquad \text{or} \qquad \left|\underline{TF} - \lambda\underline{E}\right| = 0 \tag{B3}$$

where $\underline{\Lambda}$ is a diagonal matrix of eigenvalues $\lambda_1, \lambda_2, \lambda_3, \ldots, \lambda_{3N}$ ($\lambda_i = 4\pi^2 \nu_i{}^2$, where ν_i is the frequency of the periodic motion; $\nu_i = 0$ for translation and rotation and positive for vibration), and $\underline{L}$ is the eigenvector matrix, whose i^{th} column elements $L_{1i}, L_{2i}, L_{3i}, \ldots, L_{3N,i}$, taken three at a time, give the relative atomic displacements from equilibrium.

Taking determinants of both sides of the matrix eq. (B3):

$$\det(\underline{T})\cdot\det(\underline{F}) = \det(\underline{\Lambda}) = \prod_{i=1}^{3N}\lambda_i \tag{B4}$$

Since, according to the Born-Oppenheimer approximation, the potential energy V and hence its derivatives ($f_{ij} = \partial^2 V/\partial q_i \partial q_j$) are independent of isotopic composition, the eigenvalues λ_i of a molecule are related to those of its isotopic isomer, $^*\lambda_i$ by:

$$\det(\underline{t})/\det(^*\underline{t}) = \prod_{i=1}^{3N}(\lambda_i/^*\lambda_i) \tag{B5}$$

Suppose now that the equations of motion were formulated in Cartesian coordinates. If x_{1i} and p_{1i} are the x and p_x coordinate and conjugate momentum for atom i, and subscripts 2 and 3 imply the same quantities for y and z directions, the kinetic energy is:

$$T = (1/2m_i)\sum_{i=1}^{N}(p_{1i}{}^2 + p_{2i}{}^2 + p_{3i}{}^2) \tag{B6}$$

Hence $\underline{T} = \text{diagonal}\{1/m_1, 1/m_1, 1/m_1, \ldots, 1/m_N, 1/m_N, 1/m_N\}$, and eq. B5 yields:

$$\prod_{i=1}^{N}(m_i/^*m_i)^{-3} = \prod_{i=1}^{3N}(\lambda_i/^*\lambda_i) \tag{B7}$$

Now suppose that a coordinate system is employed which translates with the center of mass of the molecule and which rotates with the molecule such that the axes coincide with the principal axes of rotation of the equilibrium configuration (the actual condition of "rotating with the molecule" such as to effect a separation of rotational and vibrational motions is given by the Eckart conditions,[142] which correspond closely to the above statement). Wilson, Decius and Cross[143] have shown that in such a coordinate system, if the (usually small) Coriolis coupling terms are neglected,

$$2T = P_{CM}^2/M + \sum_{i=1}^{3} (L_i^2/I_i) + T_{vib} \qquad (B8)$$

where M is the molecular mass, P_{CM} the momentum of the center of mass, L_i the ith component of the rotational angular momentum, I_i is the ith principal moment of inertia and T_{vib} is the vibrational kinetic energy. In matrix notation,

$$\underline{T} = \text{diagonal}\{\underline{M}^{-1}, \underline{I}^{-1}, \underline{G}\} \qquad (B9)$$

where $\underline{M}^{-1} = \text{diagonal}\{1/M, 1/M, 1/M\}$, $\underline{I}^{-1} = \text{diagonal}\{1/I_1, 1/I_2, 1/I_3\}$, and $\underline{G}$ is the inverse kinetic energy matrix for vibration, as defined by Wilson et al[32]. Equation (B5) applied in this case yields:

$$(M/\!*\!M)^{-3} \cdot \prod_{i=1}^{3} (I_i/\!*\!I_i)^{-1} \cdot (\det\underline{G}/\det\!*\!\underline{G}) = \prod_{i=1}^{3N} (\lambda_i/\!*\lambda_i) \qquad (B10)$$

Wilson, Decius and Cross[32] show that the vibrational eigenvalues λ_i, i = 1,2,3,...,3N-6 (3N-5 for a linear N-atom molecule) are obtained by solution of the vibrational secular equation:

$$|\underline{GF} - \lambda\underline{E}| = 0 \qquad \text{or} \qquad \det\underline{G} \cdot \det\underline{F} = \prod_{i=1}^{3N-6} \lambda_i \qquad (B11)$$

again, since $\underline{F}$ is independent of isotopic substitution,

$$\det\underline{G}/\det{}^*\underline{G} = \prod_{i=1}^{3N-6}(\lambda_i/{}^*\lambda_i) \tag{B12}$$

Substitution of eqs. (B7) and (B12) into eq. (B10), taking the square root of each side of the resulting equation, and substituting $(\lambda_i)^{1/2} = 2\pi\nu_i$ and $u_i = h\nu_i/k_bT$ yields the desired Teller-Redlich Product Rule:

$$\prod_{i=1}^{N}(m_i/{}^*m_i)^{3/2}\cdot\prod_{i=1}^{3N-6}(u_i/{}^*u_i) = (M/{}^*M)^{3/2}\cdot\prod_{i=1}^{3}(I_i/{}^*I_i)^{1/2} \tag{B13}$$

which is then seen to be valid within the harmonic approximation and for neglect of Coriolis coupling of rotational and vibrational states, both of which are good approximations for most molecules at ordinary temperatures.

REFERENCES

1 L. Melander and W.H. Saunders, Jr., Reaction Rates of Isotopic Molecules, John Wiley, New York, 1980.

2 T.Ando, S.-G. Kim, K. Matsuda, H. Yamataka, Y. Yukawa, A. Fry, D.E. Lewis,. L.B. Sims and J.C. Wilson, J. Amer. Chem. Soc., 103 (1981) 3505.

3 V.F. Raaen, G.A. Ropp and H.P. Raaen, Carbon-14, McGraw-Hill, New York, 1968.

4 a) H. Kwart and J. Stanulonis, J. Amer. Chem. Soc., 98 (1976) 4009.

   b) R.M. Caprioli, T.F. Fies and M.S. Story, Anal. Chem. 46 (1974) 453A.

   c) A. Weissberger and B. Rossiter, Techniques of Chemistry, John Wiley, New York, Vol. I, Part IV, 1975, pp. 206-207.

   d) W. Reimschussel and P. Paneth, Org. Mass. Spec., 15 (1980) 302.

   e) R. Murphy, Finnigan Spectra, 3, 1973.

5 J. Bigeleisen and M. Wolfsberg, Adv. Chem. Phys. I, Interscience, New York, 1958.

6 A. Fry, Pure Appl. Chem. 8 (1964) 409.

7 L.B. Sims, A. Fry, D.E. Lewis and L.T. Netherton in Synthesis and Applications of Isotopically Labeled Compounds, W.P. Duncan and A.B. Susan, Eds., Elsevier, Amsterdam, 1982, 261-266.

8 Examples include the Quantum Chemistry Program Exchange (QCPE), Department of Chemistry, Indiana University, Bloomington, IN, USA 47401 and the National Resource for Computations in Chemistry (NRCC), Lawrence Berkeley Laboratory, Berkeley, CA, USA.

9 H.S. Johnston, Gas Phase Reaction Rate Theory, Ronald Press, New York, 1966.

10 E. Buncel and C.C. Lee, Eds., Isotopes in Organic Chemistry, Elsevier, Amsterdam, Vol. 1, 1975 and subsequent volumes.

11 C.J. Collins and N.S. Bowman, Eds., Isotope Effects in Chemical Reactions, ACS Monograph No. 167, Van Nostrand Reinhold, New York, 1970.

12 J.A. Elvidge and J.R. Jones, Eds., Isotopes: Essential Chemistry and Applications, The Chemical Society (London), 1979.

13 W.W. Cleland, M.H. O'Leary and D.B. Northrop, Eds., Isotope Effects on Enzyme-Catalyzed Reactions, University Park Press, Baltimore, 1977.

14 P.A. Rock, Ed., Isotopes and Chemical Principles, ACS Symposium Series, No.11, American Chemical Society, Washington, D.C., 1974.

15 W. Spindel,Ed., Isotope Effects in Chemical Processes, ACS Advances in Chemistry Series, No. 89, American Chemical Society, Washington, D.C., 1967.

16 W.H. Saunders, Jr. and A.F. Cockerill, Mechanisms of Elimination Reactions, John Wiley, New York, 1973.

17 J.F. Duncan and G.B. Cook, Isotopes in Chemistry, Clarendon Press, Oxford, 1968.

18 N. Davidson, Statistical Mechanics, McGraw-Hill, New York, 1962, Chapt. 7.

19 a) S. Glasstone, K.J. Laidler and H. Eyring, The Theory of Rate Processes, McGraw-Hill, New York, 1941.
   b) D.G. Truhlar and B.G. Garrett, Accts. Chem. Res., 13 (1980) 440.

20 W.A. Van Hook, Chapt. 1 in Ref. 11.

21 G. Herzberg, Molecular Spectra and Molecular Structure III. Electronic Spectra and Electronic Structure of Polyatomic Molecules, Van Nostrand Reinhold, New York, 1966, pp 181–183.

22 a) For discussions of the temperature dependence of isotope effects, see M.J. Stern, W. Spindel, E.U. Monse, J. Chem. Phys., 48 (1968) 2908.
   b) P.E. Yankwich and W.E. Buddenbaum, J. Phys. Chem., 71 (1967) 1185.
   For temperature dependence of H/D isotope effects, see:
   c) H. Kwart, Accts. Chem. Res., 15 (1982) 401.
   d) A.A. Vitale and J. San Filippo, Jr., J. Amer. Chem. Soc., 104 (1982) 27, and references contained therein.

23 M.J. Stern and M. Wolfsberg, J. Chem. Phys., 45 (1966) 2618.

24 M.J. Stern and M. Wolfsberg, J. Pharm. Soc., 54 (1965) 849.

25 O. Redlich, Z. Phys. Chem., B28 (1935) 371.

26 J. Bigeleisen and M. Goeppert-Mayer, J. Chem. Phys., 15 (1947) 261.

27 J. Bigeleisen, J. Chem. Phys., 34 (1961) 1485; J. Chim. Phys., 60 (1963) 35.

28 M. Wolfsberg, J. Chim. Phys., 60 (1963) 15.

29 E.U. Monse, L. N. Kauder, and W. Spindel, J. Chem. Phys., 41 (1964) 3898.

30 W.E. Buddenbaum and V.J. Shiner, Jr., Chapt. 1 in Ref. 13.

31 S.R. Hartshorn and V.J. Shiner, Jr., J. Amer. Chem. Soc., 94 (1972) 9002.

32 E.B. Wilson, Jr., J.C. Decius and P.C. Cross, Molecular Vibrations, McGraw-Hill, New York, 1955.

33 H. Margenau and G.M. Murphy, The Mathematics of Physics and Chemistry, Van Nostrand, New York, 2nd Ed., 1956, pp 317–318.

34 G.W. Burton, L.B. Sims, J.C. Wilson, and A. Fry, J. Amer. Chem. Soc., 99 (1977) 3371.

35 I.H. Williams, Chem. Phys. Lett., 88 (1982) 462.

36 L.H. Jones and B.I. Swanson, Accts. Chem. Res., 1 (1976) 128; B.I. Swanson, J. Amer. Chem. Soc. 98 (1976) 3067.

37 D.B. Cook and J. McKenna, J. Chem. Soc., Perkin Trans. 2 (1974) 1223.

38 For tabulations of force constants see, for example:
   a) Ref. 32, pp. 175–176.
   b) G. Herzberg, Molecular Spectra and Molecular Structure, Part II: Infrared and Raman Spectra of Polyatomic

Molecules, Van Nostrand, Princeton, 1945.

c) R.G. Snyder and J.H. Schachtschneider, Spectrochim. Acta, 21 (1965) 169; J. Mol. Spectrosc., 30 (1969) 290.

d) Ref. 31.

e) C.D. Chalk, B.G. Hutley, J. McKenna, L.B. Sims, and I.H. Williams, J. Amer. Chem. Soc., 103 (1981) 260.

39 a) W.D. Gwinn, J. Chem. Phys., 55 (1971) 477.

b) H.L. Sellers, L.B. Sims, L. Schafer and D.E. Lewis, QCPE Program No. 339, Indiana University, 1977 (see Ref. 8); J. Mol. Struct., 41 (1977) 149.

c) L.B. Sims, G.W. Burton and D.E. Lewis, QCPE No. 337, 1977.

40 J.H. Schachtschneider and R.G. Snyder, Spectrochim. Acta, 19 (1963) 117.

41 M.J. Stern and M. Wolfsberg, J. Chem. Phys., 45 (1966) 4105.

42 L.T. Netherton, Ph.D. Dissertation, University of Arkansas, Fayetteville, USA, 1977.

43 L.B. Sims, A. Fry, L.T. Netherton, J.C. Wilson, K.D. Reppond and S.W. Crook, J. Amer. Chem. Soc., 95 (1972) 1364.

44 For structural parameters see, for example, Tables of Inter-atomic Distances and Configurations in Molecules and Ions, Chem. Soc. Spec. Publ., Nos. 11, 1958; 18, 1965.

45 L. Pauling, J. Amer. Chem. Soc., 69 (1947) 542.

46 J.C. Wilson, Ph.D. Dissertation, University of Arkansas, Fayetteville, USA, 1975.

47 C.A. Coulson, Valence, Oxford University Press, 2nd Ed., 1961, p. 269.

48 Ref. 47, p. 243. See also J.H. Van Vleck and A. Sherman, Rev. Mod. Phys., 7 (1935) 192.

49 a) D.K. Bohme and G.I. Mackay, J. Amer, Chem. Soc., 103 (1981) 978.

b) S.W. Benson and P.S. Nangia, J. Amer. Chem. Soc., 102 (1980) 2843.

50 R.A. More O'Ferrall, J. Chem. Soc. B1970, 274.

51 W.v.E. Doering and H.H. Zeiss, J. Amer. Chem. Soc., 75 (1953) 4733.

52 P.R. Young and W.P. Jencks, J. Amer. Chem. Soc. 101 (1979) 3288.

53 J.M. Harris, S.G. Shafer, J.R. Moffatt and A.R. Becker, J. Amer. Chem. Soc., 101 (1979) 3295.

54 R. Alder, R. Baker and J.M. Brown, Mechanisms in Organic Chemistry, John Wiley, London, 1971, p. 310.

55 R.G. Wilkins in The Study of Kinetics and Mechanisms of Reactions of Transition Metal Complexes, Allyn and Bacon, Boston, 1974, p. 185.

56 E. Buncel, C. Chuaqui and H. Wilson, J. Org. Chem., 45 (1980) 3621.

57 E. Buncel, H. Wilson and C. Chuaqui, J. Amer. Chem. Soc., 104 (1982) 4896.

58 W.P. Jencks, Chem. Rev., 72 (1972) 705.

59 G. Hammond, J. Amer. Chem. Soc., 77 (1955) 334.

60 a) E.R. Thornton, J. Amer. Chem. Soc., 89 (1967) 2915; D.A.Winey and E. R. Thornton, J. Amer. Chem. Soc., 97 (1975) 3102.

    b) J.C. Harris and J.L. Kurz, J. Amer. Chem. Soc., 91 (1970) 349.

61 A. Fry, Chem. Soc. Rev., 1 (1972) 163.

62 L.B. Sims, G.W.Burton and D.M. Brubaker, 173rd American Chemical Society National Meeting, New Orleans, Louisiana, 1977, Organic Division, Paper No. 158.

63 W.J. Albery and M.M. Kreevoy, Adv. Phys. Org. Chem., 16 (1978) 87.

64 R.M. Schowen, private communication. See also: J.Rodgers, D.A.Femec,R.L.Schowen, J. Amer. Chem. Soc., 105 (1982) 3263.

65 R.M. Badger, J. Chem. Phys., 2 (1933) 128; 3 (1934) 710.

66 D.R. Hershbach, V.W. Laurie, J. Chem. Phys., 35 (1961) 458.

67 R.A. More O'Ferrall and J. Kouba, J. Chem. Soc., B1967, 985; S. Seltzer, A. Tsolis and D. B. Denney, J. Amer. Chem. Soc., 91 (1969) 4236; N. Bergman, W.H. Saunders, Jr. and L. Melander, Acta Chim. Scand., 26 (1972) 1130.

68 L.S. Bartell, Inorg. Chem. 9 (1970) 1594; R.J. Gillespie and R. S. Nyholm, Quart. Rev. (London), 11 (1957) 339; R.R. Holmes, R.M. Dieter and J.A. Golen, Inorg. Chem., 8 (1969) 2612.

69 All vibrational analyses in Table 3 were carried out using a VFF utilizing valence (diagonal) force constants only. Force constants employed were generally the diagonalelements from standard tabulations (ref. 38) of GVFF for the same molecule, although in some cases a new vibrational analysis was undertaken to produce a GVFF with smaller off-diagonal elements than those tabulated for the molecule before abstracting the diagonal elements for the VFF. Justification for this procedure in terms of the intended use of eqn.(19), is given in the text.

70 A. Streitweiser, Jr., R.H. Jagow, R.C. Fahey and S. Suzuki, J. Amer. Chem. Soc., 80 (1958) 2326. See also V.J. Shiner, Jr., Chapt. 2 of Ref. 11 and D.E. Sunko and S. Borsic, Chapt. 3 of Ref. 11.

71 a) S.B. Kaldor and W.H. Saunders, Jr., J. Amer. Chem. Soc., 101 (1979) 7594.

    b) R.P. Bell, The Tunnel Effect in Chemistry, Chapman Hall, New York, 1980.

    c) E. Caldin and V. Gold, Proton Transfer Reactions, Chapman and Hall, New York, 1975.

    d) N.-A. Bergman, Institutionen for Organisk Kemi, Goteborgs Universitet och Chalmers Tekniska Hogskola, S-412 96 Goteborg, personal communication.

    e) see also refs. 1, 9, 11, 22c,d.

72   D.J. Miller, Rm. Subramanian and W.H. Saunders, Jr., J.
      Amer. Chem. Soc., 103 (1981) 3519 and references therein.
73   E.P. Wigner, Z. Physik. Chem., B19 (1932) 203.
74   R.P. Bell, Faraday Soc. Trans. 55 (1959) 1.
75   R.P. Bell, The Proton in Chemistry, 2nd Ed., Cornell Univ.
      Press, Ithaca, New York, 1973.
76   R.J. LeRoy, K.A. Quickert and D. LeRoy, Trans. Faraday Soc.,
      66 (1970) 2997.
77   M.J. Stern and M. Wolfsberg, Pure Appl. Chem. 8 (1964) 225;
      M. Wolfsgerg, J. Chem. Phys. 33 (1960) 21; see also J.
      Bigeleisen and M. Wolfsberg, J. Chem. Phys. 21 (1953) 1972;
      22 (1954) 1264.
78   B. Crawford, Jr. and J. Overend, J. Mol. Spectrosc. 12
      (1964) 307.
79   W.E. Buddenbaum and P.E. Yankwich, J. Phys. Chem., 71 (1967)
      3136.
80   W.E. Buddenbaum and V.J. Shiner, Jr., Can. J. Chem., 54
      (1976) 1146; V. J. Siner, Jr., Chapt. 1 in ref. 13.
81   V.J. Shiner, Jr., personal communication.
82   R.L. Schowen and P.J. Huskey, J. Amer. Chem. Soc., 105 (1983)
      5704; and private communication.
83   a) E.S. Lewis, Topics in Current Chemistry, 74 (1978) 31.
      b) E.S. Lewis in Vol. 2 of ref. 10, 1976.
      c) J.P. Klinman in Transition States of Biochemical Pro-
         cesses, R. Gandour and R. L. Schowen, Ed. Planum, New
         York, 1978, Chapt. 4.
84   a) F.H. Westheimer, Chem. Rev. 61 (1961) 265.
      b) L. Melander, Isotope Effects on Reaction Rates, Ronald
         Press, New York, 1964.
85   J. Bigeleisen, Pure Appl. Chem. 8 (1964) 217.
86   G. Klass, D.J. Underwood and J. H. Bowie, Aust. J. Chem., 34
      (1981) 507.
87   H. Eyring, J. Walter and G. Kimball, Quantum Chemistry, John
      Wiley, New York, 1944.
88   D.J. McLennan, Aust. J. Chem., 32 (1979) 1869; 32 (1979)
      1883; 35 (1982) 1045; 36 (1983) 1503; 36 (1983) 1513; 36
      (1983) 1521.
89   F.M. Hawthorne and E.S. Lewis, J. Amer. Chem. Soc., 60 (1958)
      4296.
90   R.A. More O'Ferrall, J. Chem. Soc. B1970, 785.
91   J.L. Hogg in Transition States of Biochemical Processes, R.
      D. Gandour and R. L. Schowen, Eds., Plenum, New York, 1978,
      Chapt. 5.
92   a) L.C. Kurz and C. Frieden, J. Amer. Chem. Soc., 102 (1980)
         4198.
      b) W.W. Cleland, et al, Biochemistry, 20 (1981) 1817; 19
         (1980) 4853; 20 (1981) 1797; 20 (1981) 1805.
93   C.K. Ingold, Chem. Rev. 15 (1934) 225.

94 E.D. Hughes and C.K. Ingold, J. Chem. Soc. 1935, 244; C.K. Ingold, Structure and Mechanism in Organic Chemistry, 2nd Ed., Cornell University Press, Ithaca, 1969; C.K. Ingold, Proc. Chem. Soc., London, (1962) 265.

95 S. Winstein, B. Appel, R. Baker and A. Diaz, Chem. Soc. (London) Spec. Publ., 19 (1965).

96 R.A. Sneen and J.W. Larsen, J. Amer. Chem. Soc., 91 (1969) 6031; Accts. Chem. Res. 6 (1973) 46.

97 C.J. Collins, F.J. Brown, J.F. Raaen and T. Juhlke, J. Amer. Chem. Soc. 96 (1974) 5928.

98 D.J. McLennan, J. Chem. Soc. Perkin Trans. 2 (1974) 1818; Tetrahedron, 31 (1975) 2999.

99 a) E.R. Thornton, Solvolysis Mechanisms, Ronald Press, New York, 1964.
    b) J.W. Hill and A. Fry, J. Amer, Chem. Soc., 84 (1962) 2763.
    c) E.R. Thornton, J. Amer. Chem. Soc., 89 (1967) 2915.
    d) C.G. Swain, J. Amer. Chem. Soc., 70 (1948) 1119; C.G. Swain and R.W. Eddy, J. Amer. Chem. Soc., 70 (1948) 2989.
    e) D.J. McLennan, J. Chem. Soc. Perkin Trans. 2 (1981) 1316.

100 D.M. Brubaker, Ph.D. Dissertation, University of Arkansas, 1978.

101 N. Pearson, Ph.D. Dissertation, University of Arkansas, 1970.

102 A. Fry, Chapt. 6 in ref. 11.

103 B.G. Cox and W.E. Waghorne, Chem. Soc. Rev. 9 (1980) 381.

104 J.C. Decius, J. Chem. Phys., 17 (1949) 1315.

105 A.V. Willi, Z. Naturforsch. A, 21 (1966) 1385.

106 H. Yamataka and T. Ando, Tetrahedron Lett., (1975) 1059.

107 J. McKenna, L.B. Sims and I.H. Williams, J. Amer. Chem. Soc., 103 (1981) 268.

108 H. Yamataka and T. Ando, J. Amer. Chem. Soc., 85 (1981) 2281.

109 J. Bron, Can. J. Chem., 52 (1974) 903.

110 H.S. Johnston, W.A. Bonner and D.J. Wilson, J. Chem. Phys., 26 (1957) 1002.

111 a) J. Bron and J.B. Stothers, Can. J. Chem. 47 (1969) 2506.
    b) A.J. Kresge, N.N. Lichtin, K.N. Rao and R.E. Weston, Jr., J. Amer. Chem. Soc., 87 (1965) 437.

112 M.L. Bender and G.J. Buist, J. Amer, Chem. Soc., 80 (1958) 4304.

113 a) V.J. Shiner, Jr., B.L. Murr and G. Heinemann, J. Amer. Chem. Soc., 85 (1963) 2413.
    b) L. Hakka, A. Queen and R.E. Robertson, J. Amer. Chem. Soc., 87 (1965) 161.
    c) G.J. Frisone and E.R. Thornton, J. Amer. Chem. Soc., 86 (1964) 1900; 90 (1968) 1211.

114 a) C.R. Turnquist, J.W. Taylor, E.P. Grimsrud and R.C. Williams, J. Amer. Chem. Soc., 95 (1973) 4133.
    b) R.M. Bartholomew, F. Brown and M. Lounsbury, Can. J. Chem., 32 (1954) 979.

115 R.C. Williams and J.W. Taylor, J. Amer. Chem. Soc., 95 (1973) 1710; 96 (1974) 3721.

116 J.F.Willey and J.W.Taylor, J. Amer. Chem. Soc., 102 (1980) 2387.

117 a) J.I. Brauman, W.N. Olmnsted and C.A. Lieder, J. Amer. Chem. Soc., 96 (1974) 4030; W.E. Farneth and J.I. Brauman, J. Amer. Chem. Soc. 98 (1976) 7891; W.N. Olmstead and J.I. Brauman, J. Amer. Chem. Soc., 99 (1977) 4219.

   b) A.P. Wolf, P. Schueler, R.R. Pettijohn, K.-C. To and E.P. Rack, J. Phys. Chem. 83 (1979) 1237; K.-C. To, E.P. Rack and A.P. Wolf, J. Chem. Phys. 74 (1981) 1499.

118 L.B. Sims, V.J. Shiner, Jr. and D.F. McMullen, II, unpublished results.

119 P. Pulay and L.B. Sims, research in progress.

120 a) S.Wolfe, D.J. Mitchell and H.B. Schlegel, J. Amer. Chem. Soc., 103 (1981) 7692; H.B. Schlegel, K. Mislow, F. Bernardi, A. Bottoni, Theor. Chim. Acta 44 (1977) 245.

   b) A. Dedieu and A. Veillard, J. Amer. Chem. Soc., 94 (1972) 6730.

   c) R.F.W. Bader, J.A. Duke and R.R. Messer, J. Amer, Chem. Soc., 95 (1973) 7715.

   d) F. Keil and R. Ahlrichs, J. Amer. Chem. Soc., 98 (1976) 4787.

   e) E.R. Talaty, J.J. Woods and G. Simons, Aust. J. Chem., 32 (1979) 2289.

   f) H. Fujimoto and N. Kosugi, Bull, Chem. Soc. Japan, 50 (1977) 2209.

   g) K. Ishida, K. Morokuma and A. Komornicki, J. Chem. Phys., 66 (1977) 2153; K. Morokuma, J. Amer. Chem. Soc., 104 (1982) 3732; K. Morokuma, Chapt. 4 in Frontiers of Chemistry: Plenary and Keynote Lectures, 28th IUPAC Congress, K.J. Laidler, Ed., Pergamon, New York, 1982.

121 J.F. Bunnett, Angew. Chem. 74 (1962) 731-741; Angew. Chem., Int. Ed. Engl., 1 (1962) 225-235; Surv. Prog. Chem., 5 (1969) 53.

122 a) M. Kanska and A. Fry, J. Amer. Chem. Soc., 104 (1982) 5512.

   b) A. Fry, L.B. Sims, J.R.I. Eubanks, T. Hasan, R. Kanski, F.A. Pettigrew and S. Crook, Synthesis and Application of Isotopically Labeled Compounds, W.P. Duncan and A.B. Susan, Eds., Elsevier, Amsterdam, 1983, 133-138.

   c) T. Hasan, L.B. Sims and A. Fry, J. Amer. Chem. Soc., 105 (1983) 3967.

123 D.E. Lewis, L.B. Sims, H. Yamataka and J. McKenna, J. Amer. Chem. Soc., 102 (1980) 7411.

124 W.H. Saunders, Jr., Chem. Scr., 8 (1975) 27-36; 10 (1976) 82-89; A.M. Katz and W.H. Saunders, Jr., J. Amer. Chem. Soc., 91 (1969) 4469.

125 G.W. Burton, L.B. Sims and D.J. McLennan, J. Chem. Soc., Perkin Trans. 2 (1977) 1763; 1847.

126 C.D. Chalk, J. McKenna, L.B. Sims and I.H. Williams, J. Amer. Chem. Soc., 103 (1981) 281.

127 T. Hasan, Ph.D. Dissertation, University of Arkansas, 1980.

128 J.R.I. Eubanks, Ph.D. Dissertation, University of Arkansas, 1981; D.E. Lewis, J.R.I. Eubanks and L.B. Sims, in preparation.

129 F.A. Pettigrew, Ph.D. Dissertation, University of Arkansas, 1981.

130 J.D. Roberts, Notes on Molecular Orbital Calculations, Benjamin, New York, 1982, p. 83.

131 a) M.J.S. Dewar, J. Amer. Chem. Soc., 74 (1952) 3341.
     b) L.L. Ingraham in Steric Effects in Organic Chemistry, M.S. Newman, Ed., Wiley, New York, 1956, 479.

132 T.H. Lowry and K.S. Richardson, Mechanism and Theory in Organic Chemistry, Harper and Row, 2nd Ed., New York, 1981.

133 R. Waack, L.D. McKeever and M.H. Doran, Chem. Commun. (1979) 117; V.R. Sandel and H.H.Freedman, J. Amer. Chem. Soc., 85 (1963) 2328; R. Waack and M.H. Doran, J. Amer. Chem. Soc., 85 (1963) 1651; K. Takahasi, Y. Kondo and R. Asami, J. Chem. Soc., Perkin Trans. 1 (1978) 577.

134 G.H. Cooper, J.S. Bartlett, A.M. Farid, S. Jones, D.J. Mabbott, J. McKenna, J.M. McKenna and D.G. Orchard, J. Chem. Soc. Chem. Comm. (1974) 950.

135 W.H. Saunders, Jr., private communication.

136 D.R. Wright, L.B. Sims and A. Fry, J. Amer. Chem. Soc., 105 (1983) 3714.

137 R.D. Bach, D. Andrzejewski and L.R. Dusold, J. Org. Chem., 38 (1973) 1742; W.-B. Chiao and W.H. Saunders, Jr., J. Amer. Chem. Soc., 100 (1978) 2802; H. Kwart, T.J. George, R. Louw and W. Ultee, J. Amer. Chem. Soc., 100 (1978) 3927.

138 H. Kwart and M. Brechbiel, J. Amer. Chem. Soc., 103 (1981) 4650.

139 D.R. Wright, Ph.D. Dissertation, University of Arkansas, 1984.

140 D.R. Wright, L.B. Sims and H.S.-Kermani, work in progress.

141 J.H. Keller and P.E. Yankwich, J. Amer. Chem. Soc., 96 (1974) 2303.

142 C. Eckart, Phys. Rev. 47 (1935) 552; see also ref. 32 p. 274.

143 Ref. 32, chapt. 11.

ACKNOWLEDGEMENTS
     The authors wish to thank the National Science Foundation (USA) for support of much of the work described in this report; the University of Arkansas at Fayetteville, where most of the work was carried out, for computer grants and general support; and the editors for their continued help and encouragement. Special thanks are due Professor Arthur Fry for his wise counsel, critical advice and constant good humor. The support of this project by North Carolina State University is gratefully acknowledged.